W0017284

Challenges to Tackling
Antimicrobial Resistance

Antimicrobial resistance (AMR) is a biological mechanism whereby a microorganism evolves over time to develop the ability to become resistant to antimicrobial therapies such as antibiotics. The drivers of and potential solutions to AMR are complex, often spanning multiple sectors. The internationally recognized response to AMR advocates for a 'One Health' approach, which requires policies to be developed and implemented across human, animal, and environmental health. To date, misaligned economic incentives have slowed the development of novel antimicrobials and limited efforts to reduce antimicrobial usage. However, the research which underpins the variety of policy options to tackle AMR is rapidly evolving across multiple disciplines such as human medicine, veterinary medicine, agricultural sciences, epidemiology, economics, sociology and psychology. By bringing together in one place the latest evidence and analysing the different facets of the complex problem of tackling AMR, this book offers an accessible summary for policy-makers, academics and students on the big questions around AMR policy.

This title is also available as Open Access on Cambridge Core.

MICHAEL ANDERSON is a Research Officer in Health Policy at the Department of Health Policy, London School of Economics and Political Science, and a Medical Doctor undertaking General Practice specialty training.

MICHELE CECCHINI is a Senior Health Economist, Health Division, in the Directorate for Employment, Labour and Social Affairs, Organisation for Economic Co-operation and Development.

ELIAS MOSSIALOS is Brian Abel-Smith Professor of Health Policy, Head of the Department of Health Policy at the London School of Economics and Political Science, and Co-Director of the European Observatory on Health Systems.

European Observatory on Health Systems and Policies

The volumes in this series focus on topical issues around the transformation of health systems in Europe, a process being driven by a changing environment, increasing pressures and evolving needs.

Drawing on available evidence, existing experience and conceptual thinking, these studies aim to provide both practical and policy-relevant information and lessons on how to implement change to make health systems more equitable, effective and efficient. They are designed to promote and support evidence-informed policy-making in the health sector and will be a valuable resource for all those involved in developing, assessing or analysing health systems and policies.

In addition to policy-makers, stakeholders and researchers in the field of health policy, key audiences outside the health sector will also find this series invaluable for understanding the complex choices and challenges that health systems face today.

Series Editors

JOSEP FIGUERAS Director, European Observatory on Health Systems and Policies

MARTIN MCKEE Co-Director, European Observatory on Health Systems and Policies, and Professor of European Public Health at the London School of Hygiene & Tropical Medicine

ELIAS MOSSIALOS Co-Director, European Observatory on Health Systems and Policies, and Brian Abel-Smith Professor of Health Policy, London School of Economics and Political Science

REINHARD BUSSE Co-Director, European Observatory on Health Systems and Policies, and Head of the Department of Health Care Management, Berlin University of Technology

Challenges to Tackling Antimicrobial Resistance

Economic and Policy Responses

Edited by

MICHAEL ANDERSON
London School of Economics and Political Science

MICHELE CECCHINI
OECD

ELIAS MOSSIALOS
London School of Economics and Political Science

CAMBRIDGE
UNIVERSITY PRESS

University Printing House, Cambridge CB2 8BS, United Kingdom

One Liberty Plaza, 20th Floor, New York, NY 10006, USA

477 Williamstown Road, Port Melbourne, VIC 3207, Australia

314–321, 3rd Floor, Plot 3, Splendor Forum, Jasola District Centre, New Delhi – 110025, India

79 Anson Road, #06–04/06, Singapore 079906

Cambridge University Press is part of the University of Cambridge.

It furthers the University's mission by disseminating knowledge in the pursuit of education, learning, and research at the highest international levels of excellence.

www.cambridge.org
Information on this title: www.cambridge.org/9781108799454
DOI: 10.1017/9781108864121

First published 2019

A catalogue record for this publication is available from the British Library.

ISBN 978-1-108-79945-4 Paperback

European Observatory on Health Systems and Policies

The European Observatory on Health Systems and Policies supports and promotes evidence-based health policy-making through comprehensive and rigorous analysis of health systems in Europe. It brings together a wide range of policy-makers, academics and practitioners to analyse trends in health reform, drawing on experience from across Europe to illuminate policy issues.

The European Observatory on Health Systems and Policies is a partnership hosted by the World Health Organization Regional Office for Europe, which includes the Governments of Austria, Belgium, Finland, Ireland, Norway, Slovenia, Spain, Sweden, Switzerland, the United Kingdom, and the Veneto Region of Italy; the European Commission; the World Bank; UNCAM (French National Union of Health Insurance Funds); the Health Foundation; the London School of Economics and Political Science; and the London School of Hygiene & Tropical Medicine. The Observatory has a secretariat in Brussels and it has hubs in London (at LSE and LSHTM) and at the Berlin University of Technology.

The Organisation for Economic Co-operation and Development

The OECD, which traces its roots to the Marshall Plan, groups 36 member countries committed to democratic government and the market economy. It provides a forum where governments can compare and exchange policy experiences, identify good practices and promote decisions and recommendations. Dialogue, consensus and peer review are at the very heart of OECD. The OECD continues to actively engage with countries beyond the member states, by working closely with the major emerging economies.

The principle aim of the Organisation is to promote policies for sustainable economic growth and employment, and rising standards of living. The OECD is one of the world's most reliable sources of comparable statistical, economic and social data in a broad range of public policy areas, including health, agriculture, development co-operation, education, employment, environment, taxation and trade, science, technology and industry.

By publishing robust measures of comparative health system performance, identifying and sharing good practices across our member and partner countries, and responding to country-specific demands for tailored analyses and recommendations on particular policy problems, OECD helps countries develop policies for better and healthier lives.

Contents

Foreword   *page ix*

Acknowledgements   *x*

List of figures   *xi*

List of tables   *xiii*

List of boxes   *xiv*

List of abbreviations   *xv*

List of Contributors   *xix*

1   Introduction   1

    Michael Anderson, Anuja Chatterjee, Charles Clift, Elias Mossialos

2   The health and economic burden of antimicrobial resistance   23

    Driss Ait Ouakrim, Alessandro Cassini, Michele Cecchini,
    Diamantis Plachouras

3   Tackling antimicrobial resistance in the community   45

    Sarah Tonkin-Crine, Lucy Abel, Oliver van Hecke, Kay Wang,
    Chris Butler

4   Tackling antimicrobial resistance in the hospital sector   71

    Rasmus Leistner, Inge Gyssens

5   Tackling antimicrobial resistance in the food and livestock
    sector   99

    Jeroen Dewulf, Susanna Sternberg-Lewerin, Michael Ryan

6   Fostering R&D of novel antibiotics and other technologies
    to prevent and treat infection   125

    Matthew Renwick, Elias Mossialos

7   Ensuring innovation for diagnostics for bacterial infection
    to combat antimicrobial resistance   155
    Rosanna W. Peeling, Debrah Boeras, John Nkengasong

8   The role of vaccines in combating antimicrobial resistance   181
    Mark Jit, Ben Cooper

9   The role of civil society in tackling antimicrobial resistance   207
    Anthony D. So, Reshma Ramachandran

Index   *241*

Foreword

Antimicrobial resistance (AMR) kills, increases the costs of our health care, damages trade and economies, and threatens health security. The most recent studies show that in the European Union (EU) over 33 000 people die every year due to antibiotic-resistant bacteria; of which, 75% of these deaths are caused by health care-associated infections. This is a burden that is comparable to that of tuberculosis, HIV/AIDS and influenza combined. While the use of antibiotics is slowly declining in the EU, 20% of patients are still wrongly taking them to fight a flu or cold.

The good news is there is something we can do about it. The EU is determined to contribute to this endeavour. The EU's One Health Action Plan was adopted to set out the European Commission's objectives for tackling AMR. The One Health term emphasizes that human and animal health are interconnected, together with the environment. AMR needs to be tackled not only in human health, but also in animal health and environmental policies. The EU's One Health Action Plan sets the milestones to boost research, development and innovation on AMR, shapes the global agenda and makes the EU a best practice region. Fighting AMR will not only lead to better health for European citizens, but will also benefit our economies.

This book offers all those who have a role to play a concrete, useful, and in-depth look at the health and economic burden of AMR, ways to tackle and combat AMR in all different sectors, as well as the role of vaccines and civil society.

Everyone is responsible for addressing this threat and has a part to play. So let us fight AMR together.

ANNE BUCHER
Director-General for Health and Food Safety,
European Commission

Acknowledgements

This publication arises from work funded by the European Union as part of its response to the challenge of antimicrobial resistance. The European Union provided support to the European Observatory on Health Systems and Policies so that it could review the growing evidence on policy options to combat AMR across a multitude of contexts. The Observatory worked with the London School of Economics and Political Science (LSE) and the Organisation for Economic Co-operation and Development (OECD).

We are very grateful to the authors for producing their chapters, responding to comments, and amending their chapters to reflect the evidence in other chapters. In addition, the input of LSE research assistants Nishali Patel and Sarah Averi Albala was essential while coordinating the earlier stages of chapter development. We are also grateful to the members of the Antibiotic Resistance Project from the Pew Charitable Trust who undertook an independent review of our book during the later stages of development which resulted in many useful comments and revisions.

We also extend our thanks to our colleagues at the European Observatory: Suszy Lessof, who offered a crucial steering hand in coordinating the work; and Jonathan North, Caroline White and Celine Demaret who were all involved in sharing their experiences with previous volumes and then patiently processing the final manuscript for publication.

Figures

1.1 Percentage of invasive isolates tested resistant to selected
   antibiotics for *Escherichia coli* and *Klebsiella pneumoniae*
   reported from European countries in 2017   *page* 3
1.2 Cross country comparison of patterns of *Escherichia coli* and
   Klebsiella pneumoniae resistant to third-generation
   cephalosporins   5
2.1 Relative risk of 30-day mortality of patients with resistant
   infections relative to those with susceptible infections   24
2.2 Cost of hospitalization for patients with *Escherichia coli*
   antibiotic-resistant infection and underlying drivers   34
2.3 Projected working-age population loss in OECD countries
   per year relative to 0% resistance, 2020–2050   36
4.1 Relationship between the number of hospital-acquired
   infections and investments in infection control   75
5.1 Summary of the pathways of transmission of resistant
   bacteria between animals, humans and the environment   104
5.2 Different routes for exchange of resistant bacteria or genes
   from animals to humans and vice versa   112
6.1 Number of new classes of antibiotic discovered or patented
   each decade   126
6.2 The number of antibiotics in clinical development possibly
   active against WHO PPL pathogens (2017) and the
   number of alternative therapies to antibiotics in clinical
   development (2017)   128
6.3 Framework for developing a holistic incentive package for
   antibiotic development   134
6.4 Continuum of incentivization across the antibiotic value chain 142
8.1 Ways in which vaccines may reduce antimicrobial resistance   183

9.1 Systems diagram of the challenge of antimicrobial
resistance   211
9.2 "A Fair Shot" pictograph by the Médecins sans Frontières
Access Campaign   219
9.3 Dutch Minister of Health, Welfare and Sport, Edith Schippers,
poses for photo with US Public Interest Research Group at
2016 UN General Assembly   227
9.4 Book on microbes by children for children from ReAct Latin
America   231

Tables

2.1 Estimated yearly human burden of infections due to the selected antibiotic-resistant bacteria in EU Member States, Iceland and Norway in 2007   *page* 29

3.1 Community behaviour change interventions to target antimicrobial resistance   48

3.2 A summary of systematic review evidence for three types of community-based antimicrobial stewardship interventions   56

4.1 Costs and length of stay in days by health care-associated infection type   76

4.2 EPOC definitions of the interventions and intervention components   82

4.3 Antimicrobial stewardship objectives (145 studies), type of study design and reported outcomes   84

6.1 WHO Priority Pathogens List (PPL): Global priority list of antibiotic-resistant bacteria to guide research, discovery, and development of new antibiotics   127

6.2 Push and pull incentives for antibiotic development   133

7.1 High performing biomarkers for distinguishing between bacterial and viral infections   158

7.2 Resistant pathogens posing public health threats as prioritized by the US Centers for Disease Control and Prevention   160

7.3 WHO list of priority pathogens for R&D of antibiotics   161

7.4 Pathogen–antimicrobial combinations on which GLASS will collect data   165

8.1 Vaccine and AMR status for selected important human pathogens   188

Boxes

4.1 Two examples of local good antimicrobial stewardship
practices   *page* 87
5.1 Challenges in categorizing antibiotics as therapeutic,
metaphylactic or prophylactic   102
5.2 Examples of responses to antibiotic reduction   105
5.3 Reduction of antibiotic consumption through direct guidelines 107
5.4 Surveillance of antibiotic consumption and sales data   109
5.5 Animal to human transfer of antibiotic-resistant
strains   113
5.6 Importance of cost–benefit of biosecurity reporting   115
7.1 Summary of diagnostic innovations urgently needed to reduce
misuse of antibiotics   164
7.2 Selected examples of AMR surveillance networks   168
7.3 Summary of AMR surveillance innovation needed   170

Abbreviations

ABS	Antibiotic stewardship
AGP	Antimicrobial growth promoters
AMR	Antimicrobial resistance
AMS	Antimicrobial stewardship
AMU	Antimicrobial use
ARB	Antibiotic-resistant bacteria
ARC	Antibiotic Resistance Coalition
ASP	Antibiotic stewardship programme
BARDA	Biomedical Advanced Research and Development Authority
BSI	Bloodstream infection
CARB-X	Combating Antibiotic Resistant Bacteria Biopharmaceutical Accelerator
CCCAS	Clinician Champions in Comprehensive Antibiotic Stewardship
CDC	Centers for Disease Control
CEWG	Consultative Expert Working Group
CRE	Carbapenem-resistant Enterobacteriaceae
CRP	C-reactive Protein
CSE	Centre for Science and Environment
CSF	Cerebrospinal fluid
DALY	Disbility-adjusted life-year
DDD	Defined Daily Dose
DG-RTD	Directorate-General for Research and Innovation
DP	Delayed prescribing
DRG	Diagnosis-related group
EARSS	European Antibiotic Resistance Surveillance System
EARS-NET	European Antibiotic Resistance Surveillance Network
EC	Environmental cleaning
ECDC	European Centre for Disease Prevention and Control

EDCTP	European and Developing Countries Clinical Trial Partnership
EEA	European Economic Area
EFPIA	European Federation of Pharmaceutical Industries and Associations
EFSA	European Food Safety Authority
EIB	European Investment Bank
EMA	European Medicines Agency
EPOC	Effective Practice and Organisation of Care
ESBL	Extended spectrum beta-lactamase
ETEC	Enterotoxigenic *Escherichia coli*
EU	European Union
FAO	Food and Agricultural Organization
FDA	Food and Drug Administration
G3REC	Third-generation cephalosporin
GAMRIF	Global Antimicrobial Resistance Innovation Fund
GARDP	Global Antibiotic Research and Development Partnership
GBS	Group B Streptococcus
GDP	Gross Domestic Product
GLASS	Global Antimicrobial Resistance Surveillance System
GP	General practitioner
GRACE INTRO	Genomics to combat Resistance against Antibiotics for Community-acquired LRTI in Europe/ INternet Training for Reducing antibiOtic use
HAI	Health care-associated infection/hospital-acquired infection
HIV	Human immunodeficiency virus
HTA	Health technology assessment
ICU	Intensive-care units
ID	Infectious diseases
IMI	Innovative Medicine's Initiative
IPC	Infection prevention and control
JPIAMR	Joint Programming Initiative on Antimicrobial Resistance
KISS	Krankenhaus-Infektions-Surveillance-System
LMIC	Low- and middle-income country
LOS	Length of stay

LPAD	Limited Population Antibacterial Drug
mAb	Monoclonal antibody
MER	Market entry rewards
MRSA	Methicillin-resistant *Staphylococcus aureus*
MSF	Médecins sans Frontières
NHS	National Health Service
NIAID	National Institute of Allergy and Infectious Diseases
NIH	National Institutes for Health
NIHR	National Institute for Health Research
NRDC	Natural Resources Defense Council
OECD	Organisation for Economic Co-operation and Development
OIE	World Organisation for Animal Health
PCR	Polymerase chain reaction
PDP	Product development partnerships
PIRG	Public Interest Research Group
PMDA	Pharmaceuticals and Medical Devices Agency
POC	Point of care
POCT	Point-of-care tests
PPL	Priority Pathogens List
PPS	Point prevalence surveys
PRV	Priority review vouchers
QALY	Quality-adjusted life-years
QIDP	Qualified Infectious Diseases Products
R&D	Research and development
RCDC	Research, condition, and disease categories
RCT	Randomized controlled trial
RSV	Respiratory syncytial virus
RTI	Respiratory tract infections
SDM	Shared decision-making
SME	Small and medium enterprises
STEC	Shiga toxin-producing *Escherichia coli*
SWAB	Working Party on Antibiotic Policy
TATFAR	Transatlantic Task Force on Antimicrobial Resistance
TIPR	Transferable intellectual property rights
TPP	Target Product Profile
UNICEF	United Nations Children's Fund
UI	uncertainty interval
UPEC	Uropathogenic *Escherichia coli*

USDA	US Department of Agriculture
UTI	Urinary tract infections
VRE	Vancomycin-resistant enterococci
VTEC	Verocytotoxin-producing *Escherichia coli*
WHO	World Health Organization
WTO	World Trade Organization

Contributors

Lucy Abel: Health Economist, Nuffield Department of Primary Care Health Sciences, University of Oxford, United Kingdom.

Driss Ait Ouakrim: Health Economist and Policy Analyst, Health Division, Directorate for Employment, Labour and Social Affairs, OECD, Paris, France.

Michael Anderson: Research Officer in Health Policy at the Department of Health Policy, London School of Economics and Political Science, London, United Kingdom.

Debrah Boeras: Consultant Global Health Impact Group, Atlanta, USA, International Diagnostics Centre, London School of Hygiene and Tropical Medicine, London, United Kingdom.

Chris Butler: Professor of Primary Care, Nuffield Department of Primary Care Health Sciences, University of Oxford, United Kingdom, and Salaried General Practitioner, Cwm Taf University Health Board, Abercynnon, Wales, United Kingdom.

Alessandro Cassini: Expert Antimicrobial Resistance and Healthcare-associated Infections, Scientific Advice Coordination, Surveillance and Response Support Unit, European Centre for Disease Prevention and Control (ECDC), Stockholm, Sweden.

Michele Cecchini: Senior Health Economist, Health Division, Directorate for Employment, Labour and Social Affairs, Organisation for Economic Co-operation and Development (OECD), Paris, France.

Anuja Chatterjee: PhD student, NIHR Health Protection Research Unit in Healthcare Associated Infections and Antimicrobial Resistance, Department of Medicine, Imperial College, London, United Kingdom.

Charles Clift: Senior Consulting Fellow, Chatham House, The Royal Institute of International Affairs, London, United Kingdom.

Ben Cooper: Associate Professor, Nuffield Department of Medicine, University of Oxford, United Kingdom.

Jeroen Dewulf: Professor in Veterinary Epidemiology at the Faculty of Veterinary Medicine of Ghent University, Ghent, Belgium.

Inge Gyssens: Professor of Infectious Diseases, Department of Medicine, Radboud University Medical Center, Nijmegen, The Netherlands and Faculty of Medicine, Research group of Immunology and Biochemistry, Hasselt University, Hasselt, Belgium.

Oliver van Hecke: Clinical Research Fellow and General Practitioner, Nuffield Department of Primary Care Health Sciences, University of Oxford, United Kingdom.

Mark Jit: Professor of Vaccine Epidemiology, Department of Infectious Disease Epidemiology, London School of Hygiene & Tropical Medicine, and Principal Mathematical Modeller at the Statistics, Modelling and Economics Unit, National Infections Service, Public Health England, London, United Kingdom.

Rasmus Leistner: Infection Control and Hospital Epidemiology, Institute of Hygiene and Environmental Medicine and National Reference Centre for the Surveillance of Nosocomial Infections, Charité Universitaetsmedizin, Berlin, Germany.

Elias Mossialos: Brian Abel-Smith Professor of Health Policy, London School of Economics and Political Science, London, United Kingdom, and Co-Director of the European Observatory on Health Systems, Brussels, Belgium.

John Nkengasong: Director of Africa Centres for Disease Control and Prevention, Addis Ababa, Ethiopia.

Rosanna W. Peeling: Professor and Chair, Diagnostics Research, Director, International Diagnostics Centre, London School of Hygiene and Tropical Medicine, London, United Kingdom.

Diamantis Plachouras: Senior Expert Antimicrobial Resistance and Healthcare-associated Infections, Surveillance and Response Support Unit, European Centre for Disease Prevention and Control (ECDC), Stockholm, Sweden.

Reshma Ramachandran: Associate, IDEA (Innovation+Design Enabling Access) Initiative, Johns Hopkins Bloomberg School of Public Health, Baltimore, Maryland, USA.

Matthew Renwick: Research Associate in Health Policy and Economics, Department of Health Policy, London School of Economics and Political Science, London, United Kingdom.

Michael Ryan: Senior Agricultural Policy Analyst, Trade and Agriculture Directorate, OECD, Paris, France.

Anthony D. So: Director, IDEA (Innovation+Design Enabling Access) Initiative and Director, Strategic Policy Program, ReAct – Action on Antibiotic Resistance, Department of International Health, Johns Hopkins Bloomberg School of Public Health, Baltimore, Maryland, USA.

Susanna Sternberg-Lewerin: Professor in Epizootiology and Disease Control, Department of Biomedical Sciences & Veterinary Public Health, Swedish University of Agricultural Sciences, Uppsala, Sweden.

Sarah Tonkin-Crine: Health Psychologist, Nuffield Department of Primary Care Health Sciences, University of Oxford, United Kingdom.

Kay Wang: NIHR Postdoctoral Fellow, Nuffield Department of Primary Care Health Sciences, University of Oxford, United Kingdom.

1 | Introduction

MICHAEL ANDERSON, ANUJA CHATTERJEE, CHARLES
CLIFT, ELIAS MOSSIALOS

Over the last decade, there have been several major reports on different aspects of antimicrobial resistance (AMR) (Mossialos et al., 2010; Davies, 2013; Davies, Grant & Catchpole, 2013; O'Neill, 2016a; Renwick, Simpkin & Mossialos, 2016; OECD, 2018; Renwick & Mossialos, 2018). The purpose of compiling this book was to bring together in one place the evidence and thinking from developed countries on the different facets of the complex problem of tackling AMR for academics and policy-makers. What is the evidence on the rise of AMR and its health and economic impact? How can it be most effectively addressed in the community and in hospitals? What role is played by antimicrobial use in the food and livestock sector and what can be done about it? How can the discovery of new antibiotics be reinvigorated to replace those rendered ineffective by resistance? What needs to be done to develop new diagnostic tests so that infections can be speedily identified or ruled out and unnecessary antibiotic use avoided? Can more use be made of vaccines to tackle AMR? How have civil society movements contributed to policy development in the fight against AMR?

In this book we refer to antimicrobial resistance but for the most part the argument relates specifically to antibiotics. Antibiotics are medicines used to prevent and treat *bacterial infections*. While it has been shown that some resistant forms of bacteria predate the use of antibiotics in modern medicine, the focus of this book is antibiotic resistance which occurs when bacteria change in response to the use of these medicines. Antimicrobial resistance is a broader term, encompassing resistance to drugs to treat infections caused by other microbes as well, such as parasites (e.g. malaria), viruses (e.g. HIV) and fungi (e.g. Candida). In this book we normally refer to antibiotics unless otherwise indicated but we retain the abbreviation AMR because it is in common use. Antimicrobial resistance is a biological mechanism whereby a microorganism evolves over time to develop the ability to become resistant to antimicrobial therapies such as antibiotics. The discovery of antibiotics has been one

1

of the most significant developments for humanity over the last 70 years – a breakthrough in the treatment of communicable diseases which has also facilitated developments in other areas of medicine such as surgery, obstetrics and oncology (Holmes et al., 2016; Teillant et al., 2015). The development of AMR is intrinsic to the use of antibiotics but its growth and spread is exacerbated by their overuse and misuse. This risk has been known for a long time. Sir Alexander Fleming, the discoverer of penicillin, noted this in an interview as early as 1945:

> In such cases, the thoughtless person playing with penicillin is morally responsible for the death of the man who finally succumbs to infection with the penicillin-resistant organism. I hope this evil can be averted (New York Times, 1945).

The widespread dissemination of antibiotics in the succeeding decades only increases the relevance of his words today. Following the discovery of penicillin in 1928, many classes of antibiotics followed; a period many describe as the "golden era" of antibiotic discovery. Despite this initial expansion, in the last 30 years there has been a dearth of novel antibiotics discovered (Freire-Moran et al., 2011; Spellberg et al., 2004).

Globally, the prevalence rate of resistant bacteria has been steadily increasing. Currently, in many countries rates of resistance are particularly high in Gram-negative bacteria such as *Escherichia coli* and *Klebsiella pneumoniae*; for example, in Europe (Figure 1.1).

Increased use of a major last-line antibiotic group – carbapenems – to treat these resistant infections is creating higher selection pressure resulting in more cases of carbapenem-resistant bacteria. The increased prevalence of carbapenem-resistant bacteria is a growing problem for clinicians since there are fewer alternative treatments remaining apart from, for example, colistin. However, plasmid-mediated colistin-resistant genes have now been identified in China and 30 other countries around the globe (Yin et al., 2017; Center for Infectious Disease and Research Policy, 2017). The growing rates of AMR combined with an insufficient pipeline for antibiotic discovery has yielded concerns that we may be rapidly approaching a "post-antibiotic" era (World Health Organization, 2017b).

As a result, global recognition of the threat posed by AMR has grown, with the World Health Organization (WHO) publishing the

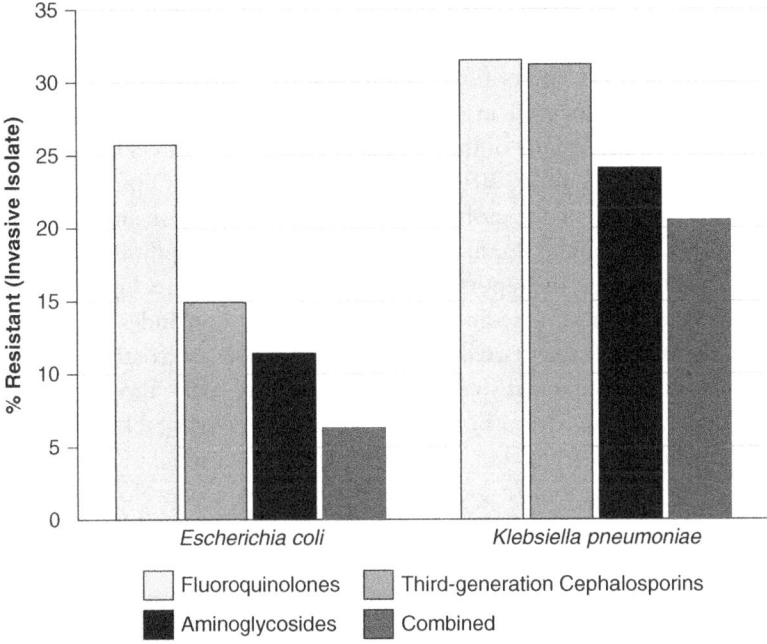

Figure 1.1 Percentage of invasive isolates tested resistant to selected antibiotics for *Escherichia coli* and *Klebsiella pneumoniae* reported from European countries in 2017

Notes: Fluoroquinolones, third-generation cephalosporins and aminoglycosides are antibiotic groups. Combined resistance refers to resistance to all three antibiotic groups. Resistance rates in this graph are the population weighted mean calculated using data reported from European Union (EU)/European Economic Area (EEA) countries.

Source: ECDC, 2018.

Global Action Plan on AMR in 2015, followed by the UN General Assembly issuing a declaration in 2016, with heads of state pledging their commitment to international cooperation to combat AMR.

In 2017, the WHO published a list of priority pathogens which outlines the antibiotic-resistant bacteria that pose the greatest threat to global public health (World Health Organization, 2017c). This list aims to guide antibiotic research and development (R&D) based on medical need as opposed to the economic factors that have traditionally directed antibiotic investment. At the top of this list, categorized as "critical", are the Gram-negative, carbapenem-resistant strains of *Acinetobacter baumannii, Pseudomonas aeruginosa*, and the Enterobacteriaceae family. In 2013, the US Centers for Disease Control and Prevention (CDC) published a US-focused urgent threats list for antibiotic resistance, which highlighted many of the same pathogens (US Centers for Disease Control and Prevention, 2013). To follow up, the WHO published an in-depth analysis of the global development pipeline for antibacterial agents (World Health Organization, 2017d). Based on optimistic clinical trial attrition rates, the report estimates that the entire pipeline could be expected to yield 10 new approvals. However, it concludes that these potential new treatments will add little to the already existing arsenal and will not be sufficient to tackle the impending AMR threat.

This recent escalation in global collective action to tackle AMR is therefore an important step, although both international and national level policy-makers must grasp this opportunity to develop national action plans which are adequately financed to address the economic and policy challenges which prevent coordinated and effective measures to contain AMR. International action is also required to incentivize the development of new antibiotics as well as other interventions (such as vaccines or diagnostics, or water and sanitation) which will be necessary to avoid the threat of a "post-antibiotic" era.

AMR in low- and middle-income countries

This book primarily focuses on evidence from high-income countries. However, it is still necessary to highlight the issue of growing AMR in low- and middle-income countries (LMICs), as AMR does not respect borders, and countries need to coordinate their actions with the rest of the global community. A review of AMR policies in LMICs has been published by the Center for Disease Dynamics, Economics & Policy (Gelband & Delahoy, 2014). In LMICs, such as India, the problem of AMR has reached critical levels (Gandra & Joshi, 2017) (Figure 1.2).

The spread of AMR is exacerbated in countries where it is common practice to buy antibiotics over the counter. The proportion of non-prescription human antibiotic use in countries outside northern Europe and North America, where the problem of AMR is greater, has been estimated at between 19% and 100% (Morgan et al., 2011). Hence, there is an increasing need to monitor this level of inappropriate antibiotic consumption and address factors common in those countries such as

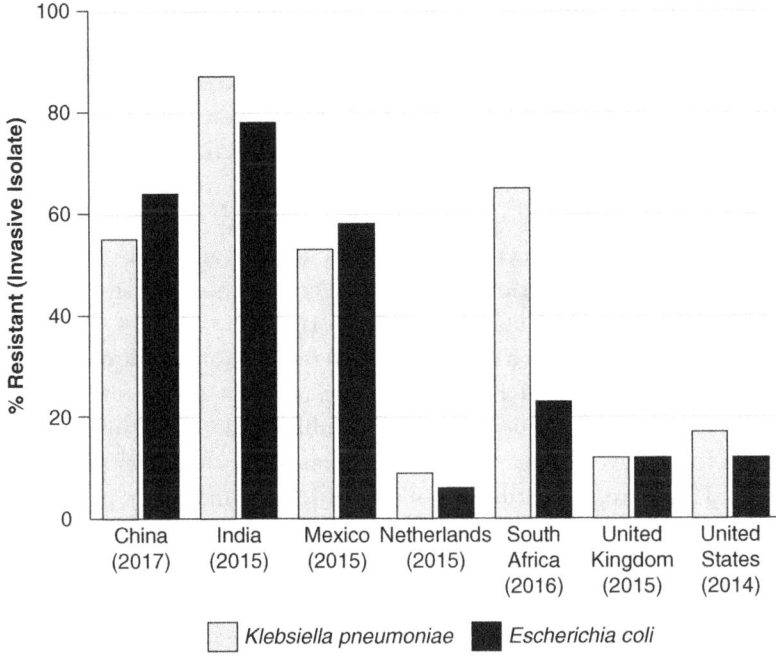

Figure 1.2 Cross country comparison of patterns of *Escherichia coli* and *Klebsiella pneumoniae* resistant to third-generation cephalosporins

Note: Years shown in brackets indicate date of most recent available data.

Source: Center for Disease Dynamics, Economics & Policy, 2018.

the misuse of antibiotics by health professionals, unskilled practitioners and laypersons; poor drug quality; unhygienic conditions accounting for spread of resistant bacteria; and inadequate surveillance (Okeke, Lamikanra & Edelman, 1999; Harbarth, 2008). Awareness of the long-term societal impact of AMR needs to be raised through global health campaigns.

Data from the ResistanceMap repository and the World Bank have proven a strong inverse association between prevalence of AMR in countries and average per capita income (Alvarez-Uria, Gandra & Laxminarayan, 2016). This is not surprising given the increasing antibiotic consumption in LMICs; in particular large emerging economies such as Brazil, Russia, India, China and South Africa (Laxminarayan, Van Boeckel & Teillant, 2015).

Many LMICs have weaker health care systems, which are associated with several factors that limit their ability to tackle AMR. These include a lack of effective infection prevention and control (IPC) practices, the affordability of second- or third-line antimicrobials, and a lack of regulation or enforcement surrounding the production of low-quality/counterfeit antimicrobials and in respect of prescribing. In addition, increasing levels of environmental contamination, a lack of reliable surveillance data and increased antibiotic use in agriculture are some of the key factors contributing to the emergence and transmission of AMR in LMICs (Laxminarayan et al., 2016).

Furthermore, there is a tension between reducing antibiotic consumption in LMICs while also ensuring appropriate access to medicines. Many LMICs have a high burden of communicable diseases, and inadequate access to effective antimicrobials, which contributes to higher mortality rates. Therefore, stewardship policies in LMICs must pay particular attention to facilitating appropriate access to effective antimicrobials alongside conservation (Laxminarayan et al., 2016).

There have been some attempts to address the problems driving AMR in LMICs. Global initiatives involving the deployment of quick and cheap rapid diagnostic tests for diagnosis of malaria, and the Affordable Medicines Facility for malaria, to incentivize the use of artemisinin to replace the use of inexpensive chloroquine treatments to which widespread resistance had developed, can be regarded as examples of success stories for tackling antimalarial resistance in LMICs (Gelband & Laxminarayan, 2015). Similarly, the Global Antibiotic Resistance Partnership aims to tackle AMR related to bacterial infections in LMICs from a national and subnational perspective. There remains an increased need for LMICs to expand their technical expertise in relation to surveillance of AMR to guide policy. There is also a need for both increased political commitment and a legal framework to guard against irresponsible antibiotic use taking a One Health perspective.

Health and economic impact of AMR

Due to AMR's intricate transmission and acquisition routes in the community and hospitals settings, it is challenging to estimate its overall aggregate burden on society. Smaller scale studies have been conducted for specific resistant bacteria types (e.g. resistant Enterobacteriaceae) within individual institutions in different countries (Stewardson et al., 2016). A number of non-comparative descriptive studies have also highlighted the issues surrounding the associated opportunity cost of AMR. However, research addressing the impact of AMR on society and the economy in terms of labour markets, trade, or tourism is limited.

Researchers investigating the health, economic, and societal burden of AMR are faced with limited data availability, resulting in uncertainty about the long-term cost and clinical efficacy outcomes for interventions (e.g. antibiotic stewardship programmes (ASPs)) currently in place to tackle AMR.

An independent AMR review commissioned by the government of the United Kingdom (O'Neill, 2016b) carried out a number of modelling studies in an attempt to estimate the burden of AMR. In the extreme scenario, where current infection rates were doubled and resistance reached 100% in all countries, it was estimated that by 2050 the deaths associated with AMR could reach 700 million with an associated cost of $14 billion (KPMG, 2014). Furthermore, the total loss of the working-age population was estimated to range from 11 to 444 million by 2050 (Taylor et al., 2014). Following the AMR review, the World Bank estimated that in a high-impact scenario, the global annual gross domestic product loss by 2030 could be around $3.4 trillion and $6.1 trillion by 2050 (Adeyi et al., 2017).

These initial modelling studies provide a bleak projection of the long-term impact of AMR on society if no action is taken. However, realistic estimates of the impact of AMR are necessary to target interventions where they are most needed. There still remains a gap in knowledge in relation to country- and setting-specific combined estimates, exploring the overall impact of these resistant strains on humans in terms of: mortality, morbidity, length of hospital stays, ineffective treatment, reduced efficacy of prophylaxis, productivity loss, or caregiver burden.

Tackling AMR in the community

Antibiotic prescribing is generally higher in the community than in hospitals. In England, about 74% of antibiotic prescriptions are in general practice compared to 11% in hospital inpatients (Public Health England, 2016). Thus, the potential for the development of resistant bacteria can occur in the community as well as hospitals. The common reasons for antibiotic prescriptions in the community include respiratory, urinary, skin, or tooth infections, with respiratory tract infections accounting for the largest share (Goossens et al., 2005; Gulliford et al., 2014; Shapiro et al., 2014).

Recent efforts to curb unnecessary antibiotic prescribing in both community and hospital settings that are explored in this book include implementation of stewardship programmes and public health campaigns. The prescribing roles of nurses and pharmacists have been cited as increasingly important due to their better compliance with protocols and prescribing guidelines (Charani et al., 2013; Wickens et al., 2013). The role of individual stakeholders and their behaviour in consulting, prescribing, dispensing, and consumption of antibiotics are seen as increasingly important in tackling AMR. A number of behaviour change interventions in the community have been trialled focusing on clinicians, patients and clinicians, and the public (Tonkin-Crine et al., 2017).

From the community or primary care perspective, recent studies have reported the clinical effectiveness in terms of reduced prescribing when interventions incorporated shared decision-making with patients (Coxeter et al., 2015), point-of-care testing (Aabenhus, Costa & Vaz-Carneiro, 2014), and delayed prescribing measures (Spurling et al., 2017). In high-income countries, these interventions appear to be effective across countries with differing health care systems but the evidence concerning effective interventions in the very different circumstances of LMICs is very limited. Social, economic, cultural and environmental factors may mean different approaches are required. In addition, most studies focus on the short-term impact of interventions and there is little evidence concerning which interventions have the greatest long-term impact.

Tackling AMR in the hospital sector

Hospitals and long-term care facilities act as reservoirs for pathogenic bacteria causing hospital-acquired infections (HAIs). Cross-transmission of pathogens between patients and health care workers via hand contact or from the hospital environment (e.g. surfaces), as well as during invasive procedures (e.g. surgery or insertion of devices), have been found to increase this risk of transmission in hospitals. These variable mechanisms of transmission highlight the need for effective IPC measures combined with stewardship programmes (Holmes et al., 2016; O'Neill, 2016b). Effective leadership and incorporation of "champions" to lead good prescribing and IPC practices, along with promotion of a positive organizational culture, have been cited as ways to tackle HAIs (Zingg et al., 2015). However, the resource constraints and overcrowding in hospitals (Clements et al., 2008) adversely affect policy compliance rates, especially in LMICs, giving rise to poor IPC measures, thus facilitating the development of resistant bacteria in hospitals (O'Neill, 2016b).

This book explores further the effectiveness and cost–effectiveness of strategies to combat AMR in the hospital sector. These strategies include IPC measures, such as contact precaution, isolation, screening, environment cleaning, and decolonization, the impact of surveillance, outbreak control measures, stewardship, and changing education curriculums at undergraduate level to influence behaviour change for nurses, pharmacists, and doctors (Pulcini & Gyssens, 2013). Recent systematic reviews and meta-analysis results have reported on the clinical effectiveness of these strategies in terms of reduction of resistant infections or prescribing (Baur et al., 2017; Karanika et al., 2016). Additional effectiveness of ASPs when combined with IPC measures have also been reported (Baur et al., 2017). However, cost–effectiveness analysis related to ASPs in hospitals is limited, but suggests ASPs may be cost-effective (Ibrahim et al., 2017).

Huge challenges remain in the effective implementation of IPC measures. Further studies are needed to estimate the excess financial costs of HAIs in order to estimate the financial scope for IPC measures and their cost–effectiveness. Too many of the current studies are of poor quality. Thus, it is recommended that action is taken to strengthen international collaboration in surveillance, that robust data are generated to provide useful information to decision-makers in hospitals and that the undergraduate curriculums of all health care professionals should include the principles of prudent prescribing.

Tackling antibiotic resistance in the food and livestock sector

The transmission of resistant microorganisms or genes may occur via direct contact between humans, or animals, or between them (including occupational, domestic or companion animal exposure). Indirect routes of transmission also exist related to travel and migration or the contamination of the food-chain from sources such as manure and soil, or due to unhygienic conditions (World Health Organization, 2017b).

In the livestock sector, antibiotic use extends beyond therapy to include use for metaphylactic, prophylactic, and growth promotion purposes. Antibiotics have often been used to boost productivity by counteracting the adverse consequences of poor and "intensive" farming conditions. A global study estimated that the volume of antimicrobials used in agriculture is expected to increase by 67% by 2030, principally as a result of increasing demand for food-producing animals and "intensive" farming in countries with growing populations, such as the USA, India, China and Brazil (Laxminarayan, Van Boeckel & Teillant, 2015).

This book explores strategies to combat AMR in the food and livestock sector such as changes to biosecurity measures to improve IPC, vaccination, husbandry practices and improved surveillance to detect emergence of infections. There are a number of economic evaluations which have reported on the effectiveness of these interventions, resulting in a reduction in antibiotic use within the livestock sector with limited repercussions on profitability (Gelaude et al., 2014; Postma et al., 2016).

The global use of antibiotics in animal production has been excessive and has resulted in selection for antibiotic resistance affecting both human and animal health. Even low doses, such as used for growth promotion, have an impact. In recent years there has been huge progress in some countries in reducing antibiotic use through improved animal management and the use of other interventions (such as vaccines or probiotics). Implementation of these measures more widely would increase this impact.

Fostering R&D of novel antibiotics and other technologies to prevent and treat infection

A number of economic, regulatory, and scientific barriers have resulted in reduced incentives to invest in new antibiotics research. Between 2007 and 2012, worldwide patent applications related to antibiotic research dropped by 34.8% (Marks & Clerk, 2015). The WHO's International Clinical Trials Registry Platform shows that there are only 182 active clinical trials that focus on bacterial infections other than tuberculosis, which is much less than 1% of the 67 000 clinical trials on noncommunicable diseases (O'Neill, 2015b).

From an economic standpoint, the return on investment in R&D on antibiotics is estimated to have, on average, a negative net present value compared to investment in other treatments, such as oncology. Due to the current stewardship practices and the existing cheaper generic alternative antibiotics, a novel class of antibiotic will be for restricted use only, reducing its revenue potential and market value. Therefore, investing in R&D of this nature is not, for the most part, attractive to private sector drug developers (Renwick, Brogan & Mossialos, 2016).

Nevertheless, the recent recognition in political and international circles of the urgent need to address AMR has resulted in a number of new funding initiatives in the USA and Europe designed to improve the pipeline for the production of new antibiotics. These include investments by existing institutions, such as through the Joint Programming Initiative on Antimicrobial Resistance, the European Commission and the Innovative Medicines Initiative in the EU. In the USA the National Institutes of Health and the Biomedical Advanced Research and Development Authority are large funders of antibiotic R&D. Two new initiatives include the Global Antibiotic Research and Development Partnership (GARDP) and Combating Antibiotic Resistant Bacteria Biopharmaceutical Accelerator (CARB-X).

While these initiatives and others have helped to boost funding on preclinical research, there remains a gap in financing the riskier and more expensive business of taking drug candidates through clinical trials. This particularly affects small and medium enterprises who are responsible for developing a large proportion of new drug candidates. Considering the widespread societal benefit associated with R&D for antibiotics, there is a need for innovative policy solutions to address this market failure. That is why consideration needs to be given to other forms of incentive such as those that offer extra rewards through the regulatory and intellectual property regimens. A combination of *pull* and *push* incentive strategies have been suggested to incentivize R&D either to boost the revenue earned from newly discovered antibiotics or to reduce the cost of R&D (Renwick, Simpkin & Mossialos, 2016). There has also been increasing support for the concept of *delinkage*; a policy tool whereby a company's return on R&D invested in a product is *delinked* from its volume-based sales (Outterson et al., 2016; O'Neill, 2015a).

Major recent reports have recommended the introduction of market entry rewards where a lump sum or phased payment would be made for the successful development of new antibiotics that meet pre-specified criteria. In its purest form, no revenue would be derived from sales so there would be no incentive for firms to maximize the consumption of the antibiotic. A less expensive version of the reward would allow some revenue to be generated from sales. It is estimated that a reward of $1–2 billion would be necessary to encourage firms to invest in R&D for new antibiotic classes (O'Neill, 2015a; Renwick, Simpkin & Mossialos, 2016; DRIVE-AB, 2018).

The impact of such a scheme would be maximized if the funding and rules were harmonized between those countries with innovative potential in antibiotics and if it were administered by a single global body established for this purpose. A possible candidate is the Global Antimicrobial Resistance Collaboration Hub established on the initiative of the G20 in 2017 (G20 Leaders' Declaration, 2017).

Ensuring innovation for diagnostics for bacterial infection

A global AMR response will require diagnostics that are affordable and accessible, can be used at the point of care (POC), and can rapidly determine antibiotic susceptibility. These tests are urgently needed to reduce inappropriate use of antibiotics, guide patient management for improved outcomes and provide much needed AMR surveillance. For any diagnostic test to be effective in primary health care settings, it needs to be simple to perform, rapid, affordable and accurate. This means providing a result in less than 15–20 minutes in order to be able to guide more targeted use of antibiotics (Okeke et al., 2011).

Traditional diagnostic tests are designed to identify pathogens in specimens taken from the patient. However, the symptoms commonly encountered where antibiotic treatment is considered can be caused by many bacterial, viral or in some cases, fungal pathogens. It would be difficult to develop a test that can identify the cause or causes of all these symptoms. As a compromise, a simple rapid test that can be used to distinguish between bacterial and viral infections would potentially be useful to inform health care providers whether a prescription for antibiotics is warranted.

The key role of diagnostic tests includes identification of target patient populations resulting in 1) reduced inappropriate antibiotic use if there were rapid-POC susceptibility tests with high sensitivity and specificity, 2) improvement in AMR surveillance data collection and timely distribution of these data and 3) reduced cost of recruiting target patient populations in costly clinical drug trials.

While the importance of diagnostics in tackling AMR is recognized, developers currently face a number of technical, policy, regulatory, financial, and implementation barriers which hamper the diagnostic innovation which is urgently needed. The UK Longitude prize initiative and a *Target Product Profile* development for a diagnostic assay to identify bacterial versus non-bacterial infection types, are some of the first steps that are being taken to reduce some of these financial and technical barriers. In order to foster the development of diagnostics, the reform of regulatory systems is necessary. Regulatory science lags far behind technological innovation, and approval processes are often lengthy, costly and not transparent (Morel et al., 2016). Access to finance is a problem – in particular because public funders do not see investing in diagnostics as having a direct impact on health outcomes. Many countries do not have national policies on diagnostic use.

A new framework for health technology assessments (HTAs) for joint review of risks and benefits by regulators and policy-makers, programme managers and subject matter experts is urgently needed, not only to facilitate a faster and more balanced regulatory review but also to accelerate implementation and policy development. Regional harmonization of a new HTA framework would also reduce duplication in clinical performance studies, reducing delays and lowering costs so that the marketed product becomes more affordable, and hence accessible.

For AMR surveillance to be effective, it is critical to: 1) understand the science and technologies needed for immediate pathogen identification to provide disease risk assessments and support global health decisions, 2) build a comprehensive network of laboratories and POC testing sites to implement quality-assured POC diagnostic services with a good laboratory–clinic interface, 3) use implementation science to understand the political, cultural, economic and behavioural context for novel diagnostic technology introduction.

As cost and funding will continue to affect innovations in diagnostics, a sound business case needs to be made to incentivize and de-risk R&D, and to finance novel diagnostic solutions for AMR. Quantifying the risk of not having diagnostics to improve the specificity of syndromic management can also encourage investments. In addition, it is important to assess the contribution of a new generation of connected diagnostics to improve the efficiency of health care systems by simplifying patient pathways, guiding appropriate use of drugs and other resources and improving patient outcomes.

Vaccines and AMR

Alongside incentivizing R&D for novel antimicrobials, alternative options exist which could reduce the demand for existing antimicrobials and in turn decrease the selective pressure of resistance with the help of vaccinations and timely diagnostic tools (O'Neill, 2016a).

Since vaccination is a population-wide preventive measure, it could directly help in reducing the infected population, resulting in increased herd immunity, reduced the transmission of infections, and lowered antibiotic use (Lipsitch & Siber, 2016). For example, vaccinations against respiratory infections offer the potential to reduce substantially the number of inappropriate antibiotic prescriptions often given for a viral infection. These vaccinations could also reduce the number of secondary bacterial infections which are sometimes associated with influenza and respiratory syncytial viral infections. It has been estimated that if universal coverage of the pneumococcal conjugate vaccine was provided to the 75 countries that currently have less than 80% coverage of this vaccine, antibiotic treatment in infected children aged less than 5 years could be halved (Laxminarayan et al., 2016). As is the case with antibiotics, the value of vaccines in combating AMR is not captured in the financial calculus when private companies take decisions on investing in vaccine R&D.

Hitherto the value of vaccines in combating AMR has not adequately been taken into account. While there are difficulties in accurately assessing this value because of the multiple pathways by which vaccines could reduce AMR, and the absence of necessary epidemiological and economic data, it is important to incorporate it when making decisions on vaccine development priorities. Three sets of pathways need to be considered. The *health systems pathway* governs the impact of vaccines on antibiotic prescribing. The *epidemiological pathways* govern the impact of vaccines on AMR directly or via reduced prescribing. This is complicated by the fact that it is not well understood how reductions in antibiotic use translate (or not) into reductions in resistance. The *economic pathways* are about how to value reductions in AMR, once determined. This could involve complex modelling of the macroeconomic effects of AMR which would involve constructing counterfactual scenarios including, for example, the cost of developing new antibiotics, or of medical procedures becoming riskier, or even impossible.

The role of the civil society in fighting AMR

The global nature of AMR has given rise to worldwide collaboration initiatives and the need for a global collective action with the increased involvement of civil society. Civil society has been recognized by the United Nations as the third sector of society along with governments and businesses. It has played a vital role in raising awareness of the repercussions of AMR. At a global policy level, civil society has high-lighted the importance of access to antibiotics (in terms of fairer prices for consumers and the public sector) and excess use of antibiotics (in terms of discouraging inappropriate or unnecessary promotion of antibiotics in LMICs by drug companies). Key contributions of civil society are the introduction of the Antibiotic Resistance Coalition and the formulation of the Antibiotic Resistance Declaration which has promoted the formulation of the Global Action Plan by WHO and the UN Political Declaration.

Just as civil society catalysed global attention over monopoly pricing of patented HIV/AIDS drugs, new civil society actors have been critical in highlighting the dearth of novel antibiotics in the R&D pipeline. Rekindling attention to AMR at WHO contributed to the policy momentum that brought the issue to the world stage. AMR, by its nature, demands an intersectoral response. Civil society organizations have successfully introduced the concept of delinkage into the policy vernacular and mobilized consumer pressure on major restaurant chains to source food animal products raised without routine use of antibiotics. This work is remarkable because of the complexity of the AMR issue, its intersectoral nature, and the fact that its victims do not readily identify themselves with this shared global health challenge. While the vision of ensuring a future free from the fear of untreatable infections is years away, the remarkable richness of the contributions that civil society has made to the policy discussions and debates over AMR offers a useful compass for future policy-making.

Conclusion: The need for global collective action

A comprehensive strategy utilizing a One Health approach targeting human, animal and environmental health is crucial to tackling AMR. To meet the five key objectives outlined within the WHO's Global

Action Plan, a multifaceted approach including antibiotic stewardship, improved global surveillance, better IPC, R&D of novel antimicrobials, diagnostic tools, and vaccines, combined with increased awareness of the threat of AMR is needed.

Individual countries can understand the need for *collective action*, whereby every country will benefit from cooperating to improve access, conservation and innovation. However, no single country is usually willing to contribute to this coordinated effort unless there are firm commitments by other countries to do the same. As a result, despite the success of individual countries in tackling domestic AMR levels (e.g. the Nordic countries and the Netherlands), there is a lack of effective internationally coordinated efforts to address the global nature of the problem. The way that global governance and global markets work can hinder or stall the search for solutions to tackling AMR on an international scale.

While there has been much discussion and proposed actions to address AMR from international organizations such as the WHO and the World Bank and in the recent declarations of the leaders of the G7 and G20, there has been little concrete progress to generate truly collective global action, although the Global Antimicrobial Resistance Collaboration Hub established in 2017 by the G20 could be an embryonic coordinating institution if supported sufficiently by a wide range of countries.

It has been argued that the national, regional and global interconnectedness of the drivers of AMR, the need to tackle simultaneously the three objectives of access, conservation and innovation, and the intersectoral actions required, make AMR a good candidate for an international legal treaty (Hoffman et al., 2015; Outterson et al., 2016). More recently a similar call has been made for an international legal agreement on AMR to be developed by a Global Steering Board and a High-Level AMR Commission (Rochford et al., 2018). However, collective action is often hampered by incentive mismatches and the competing interests of various stakeholders and institutions. A single global institution, supported by an international legal framework, could help to manage these competing interests and address the collective interest in overcoming the issues of governance, compliance, leadership and financing to achieve the shared common goal of reducing the health and economic burden of AMR.

References

Aabenhus R, Costa J, Vaz-Carneiro A (2014). Biomarkers as point-of-care tests to guide prescription of antibiotics in patients with acute respiratory infections in primary care. Cochrane Database Syst Rev. 11:CD010130.

Adeyi OO, Baris E, Jonas OB et al. (2017). Drug-resistant infections: A threat to our economic future. Washington, DC: The World Bank. (http://www.worldbank.org/en/topic/health/publication/drug-resistant-infections-a-threat-to-our-economic-future, accessed 06 September 2018).

Alvarez-Uria G, Gandra S, Laxminarayan R (2016). Poverty and prevalence of antibiotic resistance in invasive isolates. Int J Infect Dis. 52:59–61.

Baur D, Gladstone BP, Burkert F et al. (2017). Effect of antibiotic stewardship on the incidence of infection and colonisation with antibiotic-resistant bacteria and Clostridium difficile infection: a systematic review and meta-analysis. Lancet Infect Dis. 17(9):990–1001.

Center for Disease Dynamics, Economics & Policy (2018). ResistanceMap [online]. New Dehli, Washington, DC: Center for Disease Dynamics, Economics & Policy. (https://resistancemap.cddep.org/AntibioticResistance.php, accessed 06 September 2018).

Center for Infectious Disease and Research Policy (CIDRAP) (2017). New colistin resistance gene identified in China. Minnesota: Center for Infectious Disease and Research Policy. (http://www.cidrap.umn.edu/news-perspective/2017/06/new-colistin-resistance-gene-identified-china, accessed 06 September 2018).

Charani E, Castro-Sanchez E, Sevdalis N et al. (2013). Understanding the determinants of antibiotic prescribing within hospitals: the role of "prescribing etiquette". Clin Infect Dis. 57(2):188–196.

Clements A, Halton K, Graves N et al. (2008). Overcrowding and understaffing in modern health-care systems: key determinants in methicillin-resistant Staphylococcus aureus transmission. Lancet Infect Dis. 8(7):427–434.

Coxeter P, Del Mar CB, McGregor L et al. (2015). Interventions to facilitate shared decision making to address antibiotic use for acute respiratory infections in primary care. Cochrane Database Syst Rev. 11:CD010907.

Davies, SC (2013). Chief Medical Officer annual report 2011: antimicrobial resistance. London: Department of Health. (https://www.gov.uk/government/publications/chief-medical-officer-annual-report-volume-2, accessed 17 December 2018).

Davies, SC, Grant J, Catchpole M (2013). The drugs don't work: a global threat. London: Penguin Specials.

DRIVE-AB (2018). Revitalizing the antibiotic pipeline: Stimulating innovation while driving sustainable use and global access. DRIVE-AB. (http://drive-ab.eu/wp-content/uploads/2018/01/DRIVE-AB-Final-Report-Jan2018.pdf, accessed 06 September 2018).

ECDC (2018). Surveillance of antimicrobial resistance in Europe – Annual report of the European Antimicrobial Resistance Surveillance Network (EARS-Net) 2017. Stockholm: European Centre for Disease Prevention and Control. (https://www.ecdc.europa.eu/sites/portal/files/documents/AMR-surveillance-EARS-Net-2017.pdf, accessed 05 December 2018).

Freire-Moran L, Aronsson B, Manz C et al. (2011). Critical shortage of new antibiotics in development against multidrug-resistant bacteria – Time to react is now. Drug Resist Updat. 14(2):118–124.

G20 Leaders' Declaration (2017). Shaping an interconnected world. Hamburg: G20 Germany 2017. (https://www.g20.org/profiles/g20/modules/custom/g20_beverly/img/timeline/Germany/G20-leaders-declaration.pdf, accessed 06 September 2018).

Gandra S, Joshi J (2017). Scoping report on antibiotic resistance in India, 06 November 2017. Washington, D.C., New Dehli: Center for Disease Dynamics, Economics & Policy (https://cddep.org/publications/ scoping-report-antimicrobial-resistance-india/, accessed 06 September 2018).

Gelaude P, Schlepers M, Verlinden, M et al. (2014). Biocheck.UGent: A quantitative tool to measure biosecurity at broiler farms and the relationship with technical performances and antibiotic use. Poult Sci. 93:2740–2751.

Gelband H, Delahoy, M (2014). Policies to address antibiotic resistance in low- and middle-income countries, 22 September 2014. New Dehli, Washington, DC: Center for Disease Dynamics, Economics & Policy. (https://cddep.org/publications/policies_address_antibiotic_resistance_low_and_middle_income_countries/, accessed 15 December 2018).

Gelband H, Laxminarayan R (2015). Tackling antibiotic resistance at global and local scales. Trends Microbiol. 23(9):524–526.

Goossens H, Ferech M, Vander Stichele R et al. (2005). Outpatient antibiotic use in Europe and association with resistance: a cross-national database study. Lancet. 365:579–587.

Gulliford MC, Dregan A, Moore MV et al. (2014). Continued high rates of antibiotic prescribing to adults with respiratory tract infection: survey of 568 UK general practices. BMJ Open. 4:e006245.

Harbarth S (2008). Cultural and socioeconomic determinants of antibiotic use. In Gould IM, Van der Meer JWM eds. Antibiotic Policies: Fighting Resistance. Boston, MA: Springer, pp. 29–40.

Hoffman S, Caleo GM, Daulaire N et al. (2015). Strategies for achieving global collective action on antimicrobial resistance. Bull World Health Organ. 93:867–876.

Holmes AH, Moore LS, Sundsfjord A et al. (2016). Understanding the mechanisms and drivers of antibiotic resistance. Lancet. 387:176–187.

Ibrahim NH, Maruan K, Mohd Khairy HA et al. (2017). Economic evaluations on antibiotic stewardship programme: A systematic review. J Pharm Pharm Sci. 20: 397–406.

Karanika S, Paudel S, Grigoras C et al. (2016). Systematic review and meta-analysis of clinical and economic outcomes from the implementation of hospital-based antibiotic stewardship programs. Antimicrob Agents Chemother. 60(8):4840–4852.

KPMG (2014). The global economic impact of antibiotic resistance. London: KPMG. (https://home.kpmg.com/content/dam/kpmg/pdf/2014/12/amr-report-final.pdf, accessed 06 September 2018).

Laxminarayan R, Van Boeckel T, Teillant A (2015). The economic costs of withdrawing antibiotic growth promoters from the livestock sector. OECD Food, Agriculture and Fisheries Papers, No.78. Paris: OECD Publishing. (https://www.oecd-ilibrary.org/agriculture-and-food/the-economic-costs-of-withdrawing-anti-microbial-use-in-the-livestock-sector_5js64kst5wvl-en, accessed 06 September 2018).

Laxminarayan R, Paudel S, Grigoras C et al. (2016). Access to effective antimicrobials: A worldwide challenge. Lancet. 387(10014):168–175.

Lipsitch M, Siber GR (2016). How can vaccines contribute to solving the antibiotic resistance problem? mBio. 7(3):e00428–e16.

Marks & Clerk (2015). From rare to routine: Life sciences report 2015 on medicines for rare diseases, vaccines and antibiotics. London: Marks & Clerk. (https://www.marks-clerk.com/MarksClerk/media/MCMediaLib/PDF's/Reports/Life-Sciences-Report-2015-From-rare-to-routine.pdf, accessed 06 September 2018).

Morel C, McClure L, Edwards S et al. (2016) Ensuring innovation in diagnostics for bacterial infection: Implications for policy. European Observatory Health Policy Series. Copenhagen: European Observatory on Health Systems and Policies. (http://www.euro.who.int/_data/assets/pdf_file/0008/302489/Ensuring-innovation-diagnostics-bacterial-infection-en.pdf?ua=1, accessed 29 January 2019).

Morgan DJ, Okeke IN, Laxminarayan R et al. (2011). Non-prescription antibiotic use worldwide: a systematic review. Lancet Infect Dis. 11(9):692–701.

Mossialos E, Morel CM, Edwards S et al. (2010). Policies and incentives for promoting innovation in antibiotic research. Copenhagen: WHO Regional Office for Europe on behalf of the European Observatory on Health Systems and Policies. (http://www.euro.who.int/–data/assets/pdf_file/0011/120143/E94241.pdf, accessed 06 September 2018).

New York Times (1945). Penicillin finder assays its future. 26 June, p. 21.

OECD (2018). Stemming the superbug tide: just a few dollars more. Paris: OECD Publishing. (http://www.oecd.org/health/stemming-the-superbug-tide-9789264307599-en.htm, accessed 15 December 2018).

Okeke IN, Lamikanra A, Edelman R (1999). Socioeconomic and behavioral factors leading to acquired bacterial resistance to antibiotics in developing countries. Emerg Infect Dis. 5(1):18–27.

Okeke IN, Peeling RW, Goossens H et al. (2011). Diagnostics as essential tools for containing antibacterial resistance. Drug Resist Updat. 14:95–106.

O'Neill J (2015a). Securing new drugs for future generations: the pipeline of antibiotics. The Review on Antimicrobial Resistance. London: Wellcome Trust and Government of the United Kingdom. (http://amr-review.org/sites/default/files/SECURING%20NEW%20DRUGS%20 FOR%20FUTURE%20GENERATIONS%20FINAL%20WEB_0. pdf, accessed 06 September 2018).

O'Neill J (2015b). Tackling a global health crisis: initial steps. The Review on Antimicrobial Resistance. London: Wellcome Trust and Government of the United Kingdom. (https://amr-review.org/sites/default/files/ RARJ3003_Global_health_crisis_report_20.03.15_OUTLINED. pdf, accessed 06 September 2018).

O'Neill J (2016a). Tackling drug-resistant infections globally: final report and recommendations. The Review on Antimicrobial Resistance. London: Wellcome Trust and Government of the United Kingdom. (https://amr-review.org/sites/default/files/160518_Final%20paper_with%20cover.pdf, accessed 06 September 2018).

O'Neill J (2016b). Infection prevention, control and surveillance: Limiting the development and spread of drug resistance. The Review of Antimicrobial Resistance. London: Wellcome Trust and Government of the United Kingdom. (https://amr-review.org/sites/default/files/Health%20infrastructure%20and%20surveillance%20final%20 version_LR_NO%20CROPS.pdf, accessed 06 September 2018).

Outterson K, Gopinathan U, Clift C et al. (2016). Delinking investment in antibiotic research and development from sales revenues: The challenges of transforming a promising idea into reality. PLoS Med. 13(6):e1002043.

Postma M, Backhans A, Collineau L et al. (2016). Evaluation of the relationship between the biosecurity status, production parameters, herd characteristics and antibiotic usage in farrow-to-finish pig production in four EU countries. Porcine Health Manag. 2(1):9.

Public Health England (PHE) (2016). English Surveillance Programme for Antibiotic Utilisation and Resistance (ESPAUR) 2010–2015: Report 2016. London: Public Health England. (https://assets.publishing.service.gov .uk/government/uploads/system/uploads/attachment_data/file/575626/ ESPAUR_Report_2016.pdf, accessed 06 September 2018).

Pulcini C, Gyssens IC (2013). How to educate prescribers in antibiotic stewardship practices. Virulence. 4(2): 192–202.

Renwick MJ, Brogan DM, Mossialos E (2016). A systematic review and critical assessment of incentive strategies for discovery and development of novel antibiotics. J Antibiot (Tokyo). 69(2):73–88.

Renwick MJ, Simpkin V, Mossialos E (2016). Targeting innovation in antibiotic drug discovery and development: The need for a One Health – One Europe – One World Framework. Copenhagen: WHO Regional Office for Europe on behalf of the European Observatory on Health Systems and Policies. (http://www.euro.who.int/–data/assets/pdf_file/0003/315309/Targeting-innovation-antibiotic-drug-d-and-d-2016.pdf, accessed 06 September 2018).

Renwick M, Mossialos E (2018). What are the economic barriers of antibiotic R&D and how can we overcome them? Expert Opin Drug Discov. 13(10): 889–892.

Rochford C, Sridhar D, Woods N et al. (2018). Global governance of antimicrobial resistance. Lancet. 391:1976–1978.

Shapiro DJ, Hicks LA, Pavia AT et al. (2014). Antibiotic prescribing for adults in ambulatory care in the USA, 2007–09. J Antibiot Chemother. 69(1):234–240.

Spellberg B, Powers JH, Brass EP et al. (2004). Trends in antibiotic drug development: implications for the future. Clin Infect Dis. 38(9):1279–1286.

Spurling GK, Del Mar CB, Dooley L et al. (2017). Delayed antibiotic prescriptions for respiratory infections. Cochrane Database Syst Rev. 9:CD004417.

Stewardson AJ, Allignol A, Beyersmann J et al. (2016). The health and economic burden of bloodstream infections caused by antimicrobial-susceptible and non-susceptible Enterobacteriaceae and Staphylococcus aureus in European hospitals, 2010 and 2011: a multicentre retrospective cohort study. Euro Surveill. 21(33).

Taylor J, Hafner M, Yerushalmi E et al. (2014). Estimating the economic costs of antibiotic resistance: Model and results. Cambridge: RAND Corporation. (https://www.rand.org/pubs/research_reports/RR911.html, accessed 06 September 2018).

Teillant A, Gandra S, Barter D et al. (2015). Potential burden of antibiotic resistance on surgery and cancer chemotherapy antibiotic prophylaxis in the USA: a literature review and modelling study. Lancet Infect Dis. 15(12):1429–1437.

Tonkin-Crine SK, Tan PS, van Hecke O et al. (2017). Clinician-targeted interventions to influence antibiotic prescribing behaviour for acute respiratory infections in primary care: an overview of systematic reviews. Cochrane Database Syst Rev. 9:CD012252.

US Centers for Disease Control and Prevention (2013). Antibiotic resistant threats in the United States, 2013. Atlanta, Georgia: US Centers for Disease Control and Prevention. (https://www.cdc.gov/drugresistance/threat-report-2013/pdf/ar-threats-2013-508.pdf, accessed 06 September 2018).

Wickens HJ, Farrell S, Ashiru-Oredope DA et al. (2013). The increasing role of pharmacists in antibiotic stewardship in English hospitals. J Antibiot Chemother. 68(11):2675–2681.

World Health Organization (2017a). Antibacterial agents in clinical development. Geneva: World Health Organization. (http://www.who .int/medicines/news/2017/IAU_AntibacterialAgentsClinicalDevelopment_ webfinal_2017_09_19.pdf?ua=1. accessed 06 September 2018).

World Health Organization (2017b). Antibiotic Resistance (Fact sheet). WHO Media centre. Geneva: World Health Organization. (http:// www.who.int/ mediacentre/factsheets/fs194/en/, accessed 06 September 2018).

World Health Organization (2017c). Global priority list of antibiotic-resistant bacteria to guide research, discovery, and development of new antibiotics. Geneva: World Health Organization. (http://www.who.int/medicines/ publications/WHO-PPL-Short_Summary_25Feb-ET_NM_WHO.pdf, accessed 06 September 2018).

World Health Organization (2017d). Antibacterial agents in clinical development. Geneva: World Health Organization. (http://www.who .int/ medicines/news/2017/IAU_AntibacterialAgentsClinicalDevelopment_ webfinal_2017_09_19.pdf?ua=1, accessed 06 September 2018).

Yin W, Li H, Shen Y et al. (2017). Novel plasmid-mediated colistin resistance gene mcr-3 in Escherichia coli. mBio. 8 (3):e00543-e17.

Zingg W, Holmes A, Dettenkofer M et al. (2015). Hospital organisation, management, and structure for prevention of health-care-associated infection: a systematic review and expert consensus. Lancet Infect Dis. 15(2):212–224.

2 | *The health and economic burden of antimicrobial resistance*

DRISS AIT OUAKRIM, ALESSANDRO CASSINI,
MICHELE CECCHINI, DIAMANTIS PLACHOURAS

Introduction

The rising resistance to antimicrobials observed among an increasing number of microorganisms represents a direct threat to human health. Patients infected with a resistant microorganism are less likely to recover with the first antimicrobial therapy and are likely to require extra investigation and treatment – and a number of antimicrobial drugs may be needed to eradicate the infection (Cosgrove, 2006). This, in turn, results in longer hospital stays, higher morbidity and mortality for patients, as well as higher costs for the health care sector and society as a whole.

This chapter provides an overview of the health and economic burden of antimicrobial resistance (AMR). It first presents the current state of knowledge on the epidemiology of AMR and discusses the main analytical challenges in determining the current and long-term effects of resistance on populations in terms of morbidity, mortality, and length of hospital stays. In addition, a summary of the current literature on the economic impact of AMR is provided along with a detailed discussion of the characteristics and limitations of existing economic models. Finally, it identifies the main knowledge gaps and suggests avenues for future research and approaches to address them.

The effect of AMR on morbidity and mortality

AMR has profound consequences for the treatment of infections. This limitation of treatment options often makes it necessary to resort to antibiotics with a broader spectrum of action, some of which are potentially less effective or safe than narrow-spectrum antibiotics. Resistance also affects empirical treatment – where the clinician selects an antibiotic for the treatment of an infection in the absence of microbiological results – which might result in an underestimation of the risk associated with specific infections and the use of inappropriate antibiotics. For example, results from a meta-analysis showed that patients with bacteraemia caused by resistant Enterobacteriaceae are five times more likely to experience delays in receiving an effective therapy compared to patients infected by a susceptible strain (Schwaber & Carmeli, 2007). This may impair the long-term effectiveness of antibiotics, delay access to effective treatments, increase the rates of treatment failure with concomitant complications, and eventually lead to higher fatality rates. Outcome studies have consistently demonstrated increased length of stay (LOS) in hospital, greater need for surgery, and higher mortality for infections caused by resistant Gram-positive and Gram-negative bacteria (Lambert et al., 2011). Figure 2.1 illustrates the higher mortality associated with resistant infections for selected bug–drug combinations.

Another study calculated the health burden of selected antibiotic-resistant bacteria of public health importance in European Union/ European Economic Area (EU/EEA) countries expressed in

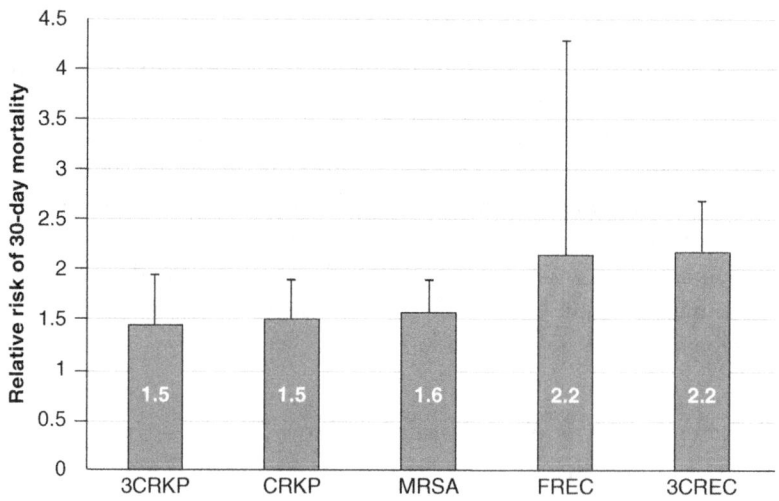

Figure 2.1 Relative risk of 30-day mortality of patients with resistant infections relative to those with susceptible infections

Notes: 3CRKP: Third-generation cephalosporin-resistant *Klebsiella pneumoniae*; CRKP: Carbapenem-resistant *Klebsiella pneumoniae*; MRSA: Methicillin-resistant *Staphylococcus aureus*; FREC: Fluoroquinolone-resistant *Escherichia coli*; 3CREC: Third-generation cephalosporin-resistant *Escherichia coli*.

Source: World Health Organization, 2014.

disabilityadjusted life-years (DALYs) (Cassini et al., 2018a). Their model was populated with estimated incidence stemming from data reported to the European Antimicrobial Resistance Surveillance Network (EARS- Net) and the European Centre for Disease Prevention and Control (ECDC) point prevalence survey of health care-associated infections and antimicrobial use in European acute care hospitals in 2011–2012 (ECDC, 2013). Moreover, data retrieved from systematic reviews of published literature provided evidence on the attributable case fatality and attributable length of stay for selected infections with antibiotic-resistant bacteria. Results showed that an estimated 671 689 (95% uncertainty intervals (UI) 583 148–763 966) infections occurred in 2015 in EU/EEA countries, accounting for 33 110 (95% UI 28 480–38 430) attributable deaths and 874 541 (768 837–989 068) DALYs. The burden of infections with bacteria resistant to antibiotics on the EU/EEA population was comparable to that of influenza, tuberculosis and HIV/AIDS combined (Cassini et al., 2018b) and, between 2007 and 2015, the burden of each of the 16 antibiotic-resistant bacteria under study has increased. The study also showed that 75% of the burden measured in DALYs was due to health care-associated infections and that this could be minimized through adequate infection prevention and control (IPC) measures, as well as antibiotic stewardship in health care settings. Finally, the contribution of various antibiotic-resistant bacteria to the overall burden varied greatly between countries, thus prevention and control strategies should be tailored to the needs of each individual country.

The impact of AMR is more serious in hospitalized patients and particularly for vulnerable groups such as immunocompromised patients (e.g. those with cancers receiving chemotherapy, having undergone organ transplantation or receiving immunosuppressive treatment) and the critically ill. These patients are also exposed to a higher risk of colonization by resistant bacteria through contact with health care delivery services, the invasive procedures their conditions often require, and frequent antibiotic treatments.

AMR also threatens the effectiveness of antibiotics as a prophylactic measure, leading to the use of broader-spectrum drugs and increasing the selection pressure for the emergence and spread of resistant strains, thereby further exacerbating the spread of AMR. Such a sequence of events jeopardizes the performance and safety of many common surgical procedures and cancer treatments that rely on effective antibiotic prophylaxis. A recent study investigated the potential health consequences of increases in antibiotic resistance for the 10 most common surgical procedures and immunosuppressing cancer chemotherapies that rely on antibiotic prophylaxis in the United States (Teillant et al., 2015). Their model showed that a 30% reduction in the efficacy of antibiotic prophylaxis for the included procedures would result on average in 120 000 additional surgical site infections and infections after chemotherapy per year (ranging from 40 000 for a 10% reduction in efficacy to 280 000 for a 70% reduction in efficacy), and 6 300 infection-related deaths (ranging from 2 100 for a 10% reduction in efficacy, to 15 000 for a 70% reduction).

The effect of AMR on the incidence of infections: replacement or addition?

In addition to the consequences on treatment effectiveness, morbidity, and mortality, AMR can also affect the number of infections in two possible ways. First, infections by resistant microorganisms can replace infections by susceptible (i.e. non-resistant) organisms through ecological replacement. In this scenario, the total number of infections remains stable as the number of infections by susceptible organisms decreases, resulting overall in increased morbidity due to resistance. Second, it has been hypothesized that the effect of some infections by *resistant* microorganisms may be additive to the burden of the same infections by *susceptible* microorganisms. In this case, the total number of infections would increase and the added burden would be the result not only of resistance but also of the additional infections. The majority of studies on methicillin-*resistant Staphylococcus aureus* (MRSA) suggest this possibility. MRSA predominantly adds to the burden of infections by methicillin-*susceptible Staphylococcus aureus*, as the incidence of infection by the latter has not decreased either in the hospital or the community, despite the increase of MRSA infections (Mostofsky, Lipsitch & Regev-Yochay, 2011). However, a combination of additive and replacement effects cannot be excluded. Additional factors may also play a role in the respective changes in incidence of infections by resistant and susceptible microorganisms. Such factors include possible differences in virulence between susceptible and resistant strains, as was shown for Panton–Valentine leucocidin-producing community-associated MRSA (Martinez-Aguilar et al., 2004). It is also likely that the predominance of additive or replacement effects differs among species or even strains of the same species.

Challenges in estimating the health burden of resistance

In a narrow sense, AMR refers to the phenomenon of a microorganism being resistant to the effect of a particular antimicrobial. It can be an intrinsic property of a microbial species (*intrinsic resistance*) or acquired by some members of the species (*acquired resistance*) through genetic or translational modifications. There is often cross-resistance to antimicrobials of the same class. For example, resistance to carbapenems implies resistance to all, or almost all, other beta-lactams. Even more alarming is the increasing frequency of microorganisms that are resistant to multiple antimicrobial classes. For example, in EU/EEA countries in 2016, 15.8% of *Klebsiella pneumoniae* isolates, on average, were resistant to beta-lactams, fluoroquinolones and aminoglycosides (ECDC, 2016).

There are a multitude of factors that determine the burden of resistance of a given pathogenic microorganism to a particular antimicrobial, including the site and severity of infections, the intrinsic resistance of particular species, and the available alternative antimicrobials that can potentially be used for the treatment of infections caused by resistant strains. In the worst-case scenario, a microorganism can be resistant to most or even all available antimicrobials, making treatment extremely challenging.

The effects of AMR are complicated by the fact that manifestations, complications, and treatment outcomes are specific to each patient. Ideally, estimation of the burden of resistance would consist of estimates for each particular microorganism and each specific antibiotic in different hosts, while at the same time accounting for multidrug-resistance phenomena. Due to the enormity of such a task, estimations of the burden of AMR currently focus on specific sites of infections, microorganisms and antimicrobial combinations.

Disease selection

From an epidemiological point of view, the selection of the pathogens and resistance combinations to assess presents an important challenge in estimating the health burden of AMR. It is also a critical step as the included pathogens and resistances will be used as indicators to determine and monitor AMR and will have a significant impact on outcome and forecasting studies, as well as on the evaluation of AMR prevention strategies.

Most efforts to estimate the burden of resistance so far have focused on specific bug–drug combinations that are common and considered clinically important due to the severity of the infections they cause and/ or the limited number of treatment options available. These combinations include: MRSA, vancomycin-resistant enterococci (VRE), extended spectrum beta-lactamase (ESBL)-producing and carbapenem-resistant Enterobacteriaceae (CRE), multidrug-resistant *Pseudomonas aeruginosa* and *Acinetobacter baumannii*, drug-resistant *Neisseria gonorrhoeae* and *Streptococcus pneumoniae*.

For example, in 2013 the US Centers for Disease Control and Prevention (CDC) provided information on the number of cases and deaths due to a selection of pathogens resistant to a number of antimicrobials (CDC, 2013). More than 2 million infections were estimated to have caused at least 23 000 deaths. In terms of mortality, the highest burden was associated with infections due to CRE, multidrug-resistant *Acinetobacter*, ESBL-producing Enterobacteriaceae, VRE, multidrug-resistant *Pseudomonas aeruginosa*, MRSA, and *Streptococcus pneumoniae*. Other infections such as drug-resistant *Candida*, drug resistant *Neisseria gonorrhoeae*, or drug-resistant *Campylobacter* are important to survey in order to follow their resistance evolution, but do not seem to cause a significant number of deaths.

In Europe in 2009, the European Centre for Disease Prevention and Control (ECDC) estimated that 386 100 resistant infections accounted for 25 100 deaths and more than 2.5 million extra hospital days (Table 2.1). MRSA had the highest incidence among AMR infections in the European context, and it was estimated that carbapenem-resistant *Pseudomonas aeruginosa* caused the highest number of deaths (ECDC/ EMA, 2009).

Data availability and sources

Estimating the incidence of AMR is challenging as the scope of most surveillance systems and data sources is to determine the proportion of resistant pathogens over the total amount of tested infections.

Table 2.1 *Estimated yearly human burden of infections due to the selected antibiotic-resistant bacteria in EU Member States, Iceland and Norway in 2007*

	No. cases of infections	No. extra deaths	No. extra hospital days
Methicillin-resistant *Staphylococcus aureus* (MRSA)	171 200	5 400	1 000
Vancomycin-resistant *Enterococcus faecium*	18 100	1 500	111 000
Penicillin-resistant *Streptococcus pneumoniae*	3 500	N/A	N/A
Third-generation cephalosporin-resistant *Escherichia coli*	32 500	5 100	358 000
Third-generation cephalosporin-resistant *Klebsiella pneumoniae*	18 900	2 900	208 000
Carbapenem-resistant *Pseudomonas aeruginosa*	141 900	10 200	809 000
Total	386 100	25 100	2 536 000

Note: N/A: Not available.

However, the number of tested samples is heterogeneous across countries and is based on a number of factors ranging from laboratory capacity, and frequency of microbiological testing to health care system organization.

Another relevant element to consider when assessing data sources is the purpose of the surveillance system or study in question. Screening, as opposed to syndromic testing, provides information on the number of positive carriers of a resistant microorganism, not necessarily suffering from an infection. For example, the ECDC's EARS-Net, similar to the Central Asian and Eastern European Surveillance of Antimicrobial Resistance (CAESAR) coordinated by the European Regional Office of the WHO, collects data only from blood and cerebrospinal fluid (CSF) for a selected number of pathogens and provides information on their susceptibility, intermediate or resistance status. The Pan American Health Organization Red Latinoamericana de Vigilancia de la Resistencia Antimicrobiana (ReLAVRA) also provide information on the proportion of resistant infections, although it is not limited to blood and CSF. For these two registries, however, differences in data quality and number of samples undermine comparability at the regional level. Finally, the recent ambitious Global Antimicrobial Resistance Surveillance System (GLASS) initiative from the WHO aims to provide information on the incidence of susceptible and resistant infections. However, the GLASS experience is too recent to allow a critical assessment of the quality of the data and it is likely that it will take time before optimal quality standards are reached globally (WHO, 2018). In a number of countries, such as Sweden, Greece, and the United Kingdom, mandatory notification of resistant bloodstream infections by particular resistant microorganisms is in place (e.g. MRSA, VRE, colistin and carbapenem resistance).

In the absence of widely available incidence data for most of the infections caused by resistant microorganisms, modelling provides another approach to estimate incidence from the available data on the proportions of resistance.

Choice of comparator

The choice of the comparator used for the estimation of the attributable case fatality/mortality and attributable LOS in hospital has an impact on the usability of results for assessing interventions. When comparing outcomes of resistant against susceptible infections, the difference measured is between treatment options for the same pathogen. On the other hand, when the comparator is no infection, the resistant infection is treated as any other infection caused by a specific pathogen (the resistant pathogen is considered distinct from the susceptible version of the same pathogen). Interventions aim at preventing AMR focus on infection control in hospitals (e.g. hand hygiene) and antibiotic stewardship. Most health care-associated infection control strategies will have an effect on all infections, irrespective of their resistance pattern. Therefore, valuable information is provided by studies on the burden of AMR using a non-infected population as comparator.

On the other hand, antibiotic stewardship aims mainly at preventing the emergence of resistance. Hence, studies describing the added burden of disease caused by resistance as opposed to that of the susceptible infections inform the effectiveness of the antibiotic stewardship intervention under investigation.

Trends in resistance rates

Although there is a general trend of increasing resistance to available antimicrobials, this trend is not uniform across European countries for bacterial species and antimicrobials. For example, EARS-Net data show decreasing rates of MRSA in several European countries (ECDC, 2017). This change has been particularly significant in countries where a national plan on infection control and antimicrobial stewardship was implemented.

By contrast, an alarming increase in resistance rates is observed in Gram-negative bacteria and especially Enterobacteriaceae. Multidrug-resistant *Escherichia coli* and *Klebsiella pneumoniae* have been increasing over the last five years. For example, EU/EEA rates of *E. coli* resistant to third-generation cephalosporins increased from 8.2% in 2009 to 13.1% in 2015. These infections are predominant in the community and are generally associated with a high and inappropriate use of antibiotics in primary care. This indicates the need to improve stewardship programmes for general practitioners.

Resistance to carbapenems, a group of antibiotics used to treat severe health care-associated infections caused by multidrug-resistant bacteria, has been spreading globally. It has led to an increase in consumption of polymyxins in several countries and, in turn, to the emergence and spread of polymyxin-resistant Enterobacteriaceae (Grundmann et al., 2017). Resistance to polymyxins, which are last-resort antimicrobials, seriously limits the treatment options for such infections. This serves as a sign of both overuse and misuse of antibiotics in hospitals, as well as poor IPC.

AMR as a negative externality in the health care sector and beyond

In economic terminology, AMR is an externality (i.e. an activity causing an effect on third parties) resulting from the use of antimicrobials to treat infections. This means that the effect of antimicrobial use in a particular patient, in terms of selection pressure and subsequent drug resistance, may not initially be felt directly by the patient or the clinician but will ultimately impact the overall welfare of other patients in the community and have adverse social and economic effects (Coast,

Smith & Millar, 1996). Determining the cost of resistance is therefore a complex task that cannot be easily performed.

The first challenge in assessing the economic burden of AMR comes from the fact that its cost is partly hidden – as neither the immediate consumer, nor the supplier of the antimicrobial, has to bear the full cost of inappropriate usage. The level of complexity increases as, similar to the health burden, estimating the economic burden requires taking into account the specificity of each microorganism in terms of single or combined resistance, treatment procedures, and associated costs.

A second challenge comes from the fact that AMR compromises the success of many medical interventions that depend on the effective treatment and prevention of infection; for example: immunosuppressive therapies, chemotherapies and surgeries. Thus, AMR can undermine the safety of hospitals and that of many interventions that require antimicrobial prophylaxis. Therefore, determining the full economic impact of resistance on health care systems requires:

1) a better understanding of the epidemiology of resistance in the context of antimicrobial prophylaxis and iatrogenic infection prevention strategies;
2) identification and measurement of the costs induced by resistance for each individual procedure.

The third challenge is that the effect of AMR goes beyond public health and has potential detrimental impacts on a number of social and economic sectors (e.g. the labour market, livestock industries, the tourism industry). Assessing the economic burden of AMR implies that its associated costs, across various sectors of the economy, should be clearly identified and measured.

Impact of AMR on the health care budget

Additional health care costs due to AMR are driven by a variety of factors such as prescription of ineffective antibiotics, delayed initiation of antimicrobial therapies, and the severity of resistant infections and the additional care they require. The treatment cost of a resistant infection has been estimated to be between $10 000 and $40 000 higher than that of a susceptible infection (Sipahi, 2008; Cohen et al., 2010; Smith & Coast, 2013; Tansarli et al., 2013; WHO, 2014). A recent modelling study conducted by the Organisation for Economic Co-operation and Development (OECD) including 33 EU and OECD countries estimated the extra health care expenditure due to AMR at around $3.5 billion per year. By 2050, the cumulative cost of AMR to the health care system of those countries is expected to reach $134 billion (OECD, 2018). The main drivers underlying this extra cost include:

- the use of second-line antibiotics (which are usually more expensive), or application of different combinations of antibiotics before identifying the most effective strategy;
- advanced laboratory tests to identify effective therapies for specific agents or imaging to monitor the development of complications associated with a given resistant infection;
- higher treatment intensity including hospitalization in the case of resistant community-acquired infections. If a patient develops a resistant infection during hospitalization, transfer to intensive care and isolation measures will substantially increase treatment cost;
- higher probabilities of undergoing surgical procedures for patients with resistant infections; these procedures may range from removal of infected tissue to amputation (Cosgrove, 2006);
- excess LOS or treatment until the infection is eradicated. This entails use of additional medical and hospital resources;
- changes in physicians' prescribing habits as they may start prescribing second-line antibiotics to treat first-line antibiotic susceptible infections if the prevalence of resistance is perceived as high (McNulty et al., 2011).

A recent study calculated the contribution of the different items to the total health care expenditure of patients with an *E. coli* bloodstream infection (Tumbarello et al., 2013) (Figure 2.2). More than half of the extra expenditure was allocated to costs associated with additional nursing and medical care, while pharmacy services (e.g. second-line therapies, broad-spectrum drugs, disposables) accounted for less than 2% of the additional costs. In some cases, the contribution of the pharmacy services, particularly in terms of second-line therapies, to the additional expenditure associated with AMR may become much larger than the estimate reported, both in absolute and relative terms. Filice and colleagues (2010) found that the costs of antibiotics to treat resistant strains of *S. aureus* were on average seven times higher than the cost of treating susceptible infections – $142 as opposed to $21.

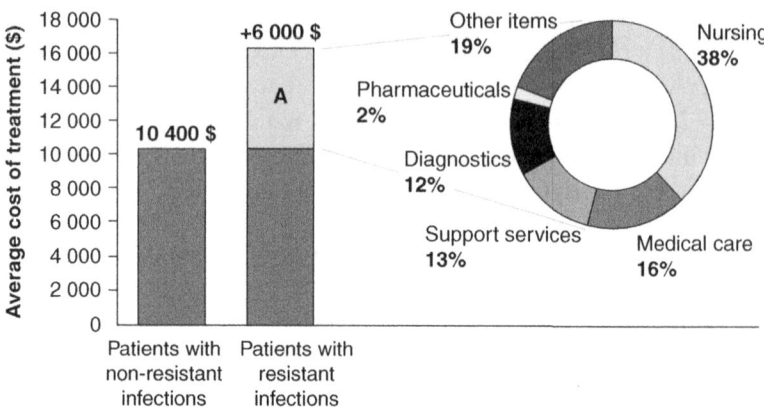

Figure 2.2 Cost of hospitalization for patients with *Escherichia coli* antibiotic-resistant infection and underlying drivers

Note: Section A reflects the additional average costs for those patients hospitalized with *E. coli-resistant* infections.

Source: OECD analysis of Tumbarello et al., 2010.

In the United States, the cost associated with treating ear infections increased by 20% (equivalent to $216 million) between 1997 and 1998 due to resistance (Sharma & Towse, 2011). The WHO estimated that the cost of treating multidrug-resistant tuberculosis in high-income countries can range from $35 000 to $41 000 per case (Fitzpatrick & Floyd, 2012). But this cost may become much higher in the case of extensively drug-resistant tuberculosis. Several reports have documented cases with treatment costs exceeding $200 000 and at least one case with a total cost close to $1 million (Chaulk & Kazandjian, 1998).

Long-term societal costs of AMR

The effect of AMR on health budgets is substantial but represents only a small fraction of the potential financial and human consequences of resistance on society. The effects of AMR on societal outcomes are determined mainly by the higher morbidity and mortality it leads to. These two factors affect the size of the labour force and labour productivity. More specifically, AMR is associated with societal costs resulting from lost income due to longer time away from work, the costs associated with ill-health and, eventually, death.

A study calculated the costs attributable to mortality and productivity loss during the extended time spent in hospitals for a cohort of US patients in 2000 and estimated the societal costs associated with resistant infections at around $38 000 per patient – more than double the medical costs (Roberts et al. 2009). This estimate does not include other potential costs incurred by the families of hospitalized persons (e.g. travel time or absence from work to care for the patient). The authors estimated that scaling up these figures to the national level would mean that the US population, in 2000, had lost about $35 billion (or about 0.35% of the national GDP) due to lost wages and premature deaths. This figure does not account for antimicrobial-resistant infections in the community. Similarly, it was calculated that, in Europe, productivity losses due to absence from work caused by AMR amounted to about €600 million in 2007 (ECDC/EMA, 2009).

At the population level, it was estimated that by 2020, the working-age population in OECD countries could be 0.6 million lower than its level in 2014 due to AMR (Taylor et al., 2014). By 2050, the total loss in people within productive age could rise to 2.1 million. Both estimates assume no increase in the level of resistance, which is an unlikely scenario, particularly if no significant action is taken against AMR. Figure 2.3 presents projections to 2050 of the potential effect of AMR on the labour force. Under two hypothetical scenarios of resistance rates of 40% and 100%, the model predicted that by 2050 the total annual number of deaths in the working-age population would reach, 4 and 10.2 million, respectively.

Macroeconomic effects of AMR

The macroeconomic effects of AMR are likely to be significant and to affect a number of sectors. As mentioned in the previous section, AMR has a detrimental effect on the labour force participation and productivity as well as on the size of the population. Both of these factors are key drivers of economic growth (Bloom, Canning & Sevilla, 2004) and several studies have attempted to provide global estimates of the economic burden of AMR by taking them into account.

In 2014, KPMG analysed the global economic impact of AMR and its potential evolution by 2050 (KPMG, 2014). The study assessed four

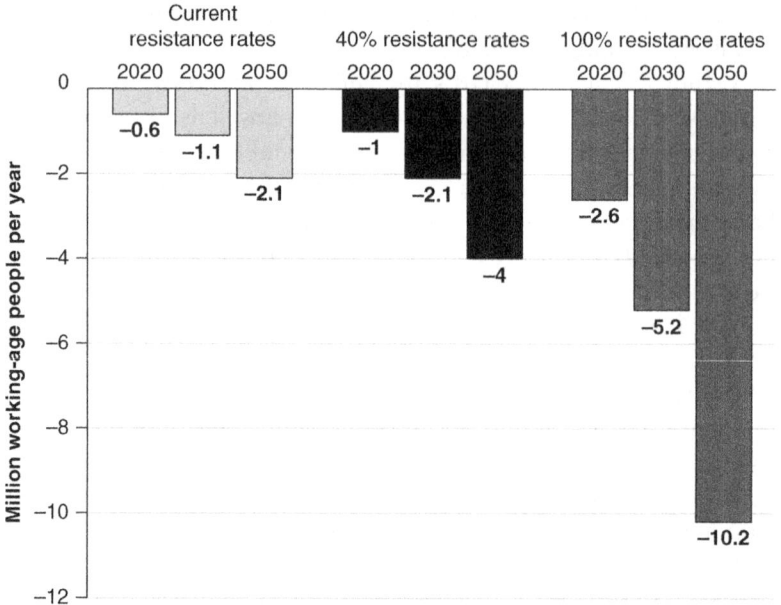

Figure 2.3 Projected working-age population loss in OECD countries per year relative to 0% resistance, 2020–2050

Source: Taylor et al., 2014.

alternative resistance scenarios for MRSA, *E. coli* and *K. pneumoniae* resistant to third-generation cephalosporins, HIV, and tuberculosis. In the most severe scenario modelled, corresponding to a doubling of current infection rates for all the infections included in the study, the authors estimated that by 2050, 700 million deaths would occur as a direct result of resistance, which would inflict a cumulative cost of over $14 trillion to the world economy.

In a similar study, RAND Europe (Taylor et al., 2014) modelled the effect of resistance on economic production through its negative impact on the labour supply. The model included the same diseases as those considered in the KPMG study, along with malaria. Different scenarios of resistance were compared to a baseline scenario of no resistance in five broad regions across the world. The model estimated that by 2050, relative to a world with no resistance, the total loss of people in productive age would range from 11 million for 5% increase in the current rates of AMR to 444 million under a 100% resistance scenario. This would correspond to a cumulative GDP loss to the global economy of between \$2.1 and \$58.9 trillion, over a period of 40 years.

Using a general equilibrium model, the World Bank (Adeyi et al., 2017) assessed the impact of AMR on global GDP and on specific components of the world economy between 2017 and 2050. Two alternative scenarios of low and high AMR impact were simulated as shocks to labour supply. In the optimistic case of low AMR impacts, the simulations estimated that, by 2050, annual global GDP would likely fall by 1.1%, relative to a base-case scenario with no AMR effects. The corresponding GDP shortfall would exceed \$1 trillion annually after 2030. In the high AMR impact scenario, the world would lose 3.8% of its annual GDP by 2050, with an annual shortfall of \$3.4 trillion by 2030. International trade and livestock production were identified as the two areas, in addition to the health care sector, that may suffer the most due to reduced productivity and reduction in sales. In certain regions, antimicrobials are largely used in the agriculture sector as an alternative to other more expensive options to allow high-density livestock production. Low- and middle-income countries are particularly exposed as livestock production and export represent important sectors of their economies.

Trade may also be affected by lower demand for animal products from consumers and limitations to imports. For example, in 2015, the negative media publicity about infections in poultry caused a 15–20% drop in sales of chickens in Norway which continued for several months (O'Dwyer, 2015). Similar events may also provoke trade disruptions, with countries imposing bans on imports following disease outbreaks. These reactions sharply reduce and disrupt economic activity, particularly in the case of diseases for which no effective cure is available (Brahmbhatt & Dutta, 2008). More broadly, it has been hypothesized that the effect of AMR may follow patterns similar to those of epidemic outbreaks developing into pandemics (Anderson, 1999; Spellberg et al., 2008). If this happened, negative effects and financial losses in the sectors such as tourism and banking could also occur (Jonas, 2013).

Conclusion

AMR is a complex phenomenon that hinders the treatment and prevention of infections and threatens the effective provision of health care. Multiple studies have demonstrated that AMR is associated with increased morbidity and mortality. However, estimating the total health burden of AMR is challenging given the number and types of resistance, microorganisms, sites of infection and hosts. The availability, quality and comparability of data are additional limiting factors.

A common feature of most existing evaluations is the limitation of their scope as they often focus on a specific infectious disease or small set of diseases. In particular, economic evaluation studies have consistently failed to consider "the bigger picture" when it comes to assessing AMR (Smith & Coast, 2013). As discussed earlier in the chapter, this is due to a large extent to the complex nature of the problem of resistance – lack of data, high parameter uncertainty, and unknown long-term consequences of interventions. Studies are easier to perform if their scope is limited to specific resistances or to the "micro" level of individual institutions (Coast et al., 2002). This has led to the somewhat paradoxical current situation, where more empirical information on the economic burden of resistance is available, but the value of that information to those in charge of designing and implementing strategies to deal with resistance is limited.

The vast majority of existing economic studies tend to consider the costs and health outcomes due to resistance without comparison. If we define economic evaluation as "the comparative analysis of alternative courses of action in terms of both their costs and consequences" (Drummond et al., 2005), most studies on the economic impact of AMR can be considered as partial economic evaluations. They provide valuable and detailed descriptive information in terms of cost and health consequences of resistance. This kind of descriptive work is important but does not provide a complete picture of the problem of AMR either in terms of costs or effects. A more comprehensive understanding of the problem requires an estimate of the "opportunity cost" associated with resistance. That is what society is missing out on by committing resources to dealing with the causes and consequences of resistance and not allocating or using those resources to do something else. To determine that opportunity cost, it is necessary to compare alternative courses of action or events in terms of both their costs and consequences. The slow progress in combating AMR is partly due to an insufficient or poor evidence base for the effectiveness and cost–effectiveness of the many existing policies across the human health and animal sectors (Dar et al., 2016). In the human health sector, the OECD has recently published a set of analyses evaluating the cost–effectiveness of selected policies to tackle AMR in different countries (OECD, 2018). The results of this work showed that investing and implementing preventive strategies – such as hand hygiene, surface cleaning and stewardship programmes – at the national level, would significantly reduce the health and economic burden of AMR and deliver high value for money. Similar evaluation work is needed to assess the effectiveness and cost–effectiveness of interventions to address AMR in the animal and environmental sectors.

References

Adeyi OO, Baris E, Jonas OB et al. (2017). Drug-resistant infections: a threat to our economic future (Vol. 2): final report. Washington, DC: World Bank Group. (http://documents.worldbank.org/curated/en/323311493396993758/final-report, accessed 06 September 2018).

Anderson RM (1999). The pandemic of antibiotic resistance. Nat Med. 5(2):147–149.

Bloom DE, Canning D, Sevilla J (2004). The effect of health on economic growth: a production function approach. World development. 32(1):1–13.

Brahmbhatt M, Dutta A (2008). On SARS type economic effects during infectious disease outbreaks. Washington, DC: World Bank Group. (http://documents.worldbank.org/curated/en/101511468028867410/ n-SARS-type-economic-effects-during-infectious-disease-outbreaks, accessed 06 September 2018).

Cassini A, Högberg LD, Plauchouras D et al. (2018a). Attributable deaths and disability-adjusted life-years caused by infections with antibiotic-resistant bacteria in the EU and the European Economic Area in 2015: a population-level modelling analysis. Lancet Infect Dis. doi: S1473-3099(18):30605-4.

Cassini A, Colzani E, Pini A et al. (2018b). Impact of infectious diseases on population health using incidence-based disability-adjusted life years (DALYs): results from the Burden of Communicable Diseases in Europe study, European Union and European Economic Area countries, 2009 to 2013. Eurosurveillance. 23(16). doi: 10.2807/1560-7917.ES.2018.23.16.17-00454.

Chaulk CP, Kazandjian VA (1998). Directly observed therapy for treatment completion of pulmonary tuberculosis: Consensus Statement of the Public Health Tuberculosis Guidelines Panel. JAMA. 279(12):943–948.

Coast J, Smith RD, Millar MR (1996). Superbugs: should antimicrobial resistance be included as a cost in economic evaluation? Health Econom. 5(3):217–226.

Coast J Smith R, Karcher AM et al. (2002). Superbugs II: how should economic evaluation be conducted for interventions which aim to contain antimicrobial resistance? Health Econom. 11(7):637–647.

Cohen B, Larson EL, Stone PW et al. (2010). Factors associated with variation in estimates of the cost of resistant infections. Med Care. 48(9):767–775.

Cosgrove SE (2006). The relationship between antimicrobial resistance and patient outcomes: mortality, length of hospital stay, and health care costs. Clin Infect Dis. 42(2):S82–89.

Coxeter P, Del Mar CB, McGregor L et al. (2015). Interventions to facilitate shared decision making to address antibiotic use for acute respiratory infections in primary care. Cochrane Database Syst Rev.11:CD010907.

Dar OA, Hasan R, Schlundt J et al. (2016). Exploring the evidence base for national and regional policy interventions to combat resistance. Lancet. 387(10015):285–295.

Drummond MF, Sculpher MJ, Claxton K et al. (2005). Methods for the economic evaluation of health care programmes. New York: Oxford University Press.

European Centre for Disease Prevention and Control (ECDC) (2013). Point prevalence survey of healthcare-associated infections and antibiotic use in European acute care hospitals, 2011–2012. Stockholm: European Centre for Disease Prevention and Control. (https://ecdc.europa.eu/sites/portal/files/media/en/publications/Publications/healthcare-associated-infections-antimicrobial-use-PPS.pdf, accessed 06 September 2018).

European Centre for Disease Prevention and Control (ECDC) (2016). Antimicrobial resistance surveillance in Europe 2010–2016. Stockholm: European Centre for Disease Prevention and Control. (https://ecdc.europa.eu/en/publications-data/antimicrobial-resistance-surveillance-europe-2016, accessed 06 September 2018).

European Centre for Disease Prevention and Control (ECDC) (2017). Antimicrobial resistance surveillance in Europe 2015. Stockholm: European Centre for Disease Prevention and Control. (https://ecdc.europa.eu/sites/portal/files/media/en/publications/Publications/antimicrobial-resistance-europe-2015.pdf, accessed 06 September 2018).

European Centre for Disease Prevention and Control/European Medicines Agency (ECDC/EMA) (2009). ECDC/EMEA Joint Technical Report: The bacterial challenge: time to react. Stockholm: European Centre for Disease Prevention and Control. (https://ecdc.europa.eu/en/publications-data/ecdcemea-joint-technical-report-bacterial-challenge-time-react, accessed 06 September 2018).

Filice GA, Nyman JA, Lexau C et al. (2010). Excess costs and utilization associated with methicillin resistance for patients with Staphylococcus aureus infection. Infect Cont Hosp Epidemiol. 31(4):365–373.

Fitzpatrick C, Floyd K (2012). A systematic review of the cost and cost effectiveness of treatment for multidrug-resistant tuberculosis. Pharmacoeconom. 30(1):63–80.

Grundmann H, Glasner C, Albiger B et al. (2017). Occurrence of carbapenemase-producing Klebsiella pneumoniae and Escherichia coli in the European survey of carbapenemase-producing Enterobacteriaceae (EuSCAPE): a prospective, multinational study. Lancet Infect Dis. 17(2):153–163.

Jonas OB (2013). Pandemic risk. Background paper for the World Development Report. Washington, DC: World Bank Group. (http://www.worldbank.org/content/dam/Worldbank/document/HDN/Health/WDR14_bp_Pandemic_Risk_Jonas.pdf, accessed 06 September 2018).

KPMG (2014). The global economic impact of antibiotic resistance. London: KPMG. (https://home.kpmg.com/content/dam/kpmg/pdf/2014/12/amr-report-final.pdf, accessed 06 September 2018).

Lambert ML, Suetens C, Savey A et al. (2011). Clinical outcomes of healthcare-associated infections and antimicrobial resistance in patients admitted to European intensive-care units: a cohort study Lancet. Infect Dis. 11(1):30–38.

Martinez-Aguilar, Avalos-Mishaan A, Hulten KG et al. (2004). Community-acquired, methicillin-resistant and methicillin-susceptible Staphylococcus aureus musculoskeletal infections in children. Pediatr Infect Dis J. 23(8):701–706.

McNulty CA, Lasseter GM, Charlett A et al. (2011). Does laboratory antibiotic susceptibility reporting influence primary care prescribing in urinary tract infection and other infections? J Antimicrob Chemother. 66(6):1396–1404.

Mostofsky E, Lipsitch M, Regev-Yochay G (2011). Is methicillin-resistant Staphylococcus aureus replacing methicillin-susceptible S. aureus? J Antimicrob Chemother. 66(10):2199–2214.

O'Dwyer G (2015). Chicken sales fall amid consumer health concerns. Global Meat News. 2 March. (https://www.globalmeatnews.com/Article/2015/03/02/Chicken-sales-fall-amid-consumer-health-concerns, accessed 06 September 2018).

OECD (2018). Stemming the superbug tide: just a few dollars more. Paris: OECD Publishing. (http://www.oecd.org/health/stemming-the-superbug-tide-9789264307599-en.htm, accessed 15 December 2018).

Roberts RR Hota B, Ahmad I et al. (2009). Hospital and societal costs of antimicrobial-resistant infections in a Chicago teaching hospital: implications for antibiotic stewardship. Clin Infect Dis. 49(8):1175– 1184.

Schwaber MJ, Carmeli Y (2007). Mortality and delay in effective therapy associated with extended-spectrum beta-lactamase production in Enterobacteriaceae bacteraemia: a systematic review and meta-analysis. J Antimicrob Chemother. 60(5):913–920.

Sharma P, Towse A (2011). New drugs to tackle antimicrobial resistance. London: Office of Health Economics. (https://www.ohe.org/publications/new-drugs-tackle-antimicrobial-resistance-analysis-eu-policy-options, accessed 06 September 2018).

Sipahi OR (2008). Economics of antibiotic resistance. Exp Rev Anti-Infect Ther. 6(4):523–539.

Smith R, Coast J (2013). The true cost of antimicrobial resistance. BMJ (Clinical Research edn). 346:f1493.

Spellberg B, Guidos R, Gilbert D et al. (2008). The epidemic of antibiotic-resistant infections: a call to action for the medical community from the Infectious Diseases Society of America. Clin Infect Dis. 46(2):155– 164.

Tansarli GS, Karageorgopoulos DE, Kapaskelis A et al. (2013). Impact of antimicrobial multidrug resistance on inpatient care cost: an evaluation of the evidence. Exp Rev Anti-Infect Ther. 11(3):321–331.

Taylor J, Hafner M, Yerushalmi E et al. (2014). Estimating the economic costs of antibiotic resistance: Model and results. Cambridge: RAND Corporation. (https://www.rand.org/pubs/research_reports/RR911. html, accessed 06 September 2018).

Teillant A, Gandra S, Barter D et al. (2015). Potential burden of antibiotic resistance on surgery and cancer chemotherapy antibiotic prophylaxis in the USA: a literature review and modelling study. Lancet Infect Dis. 15(12):1429–1437.

Tumbarello M, Spanu T, Di Bidino R et al. (2010). Costs of bloodstream infections caused by Escherichia coli and influence of extended-spectrum-beta-lactamase production and inadequate initial antibiotic therapy. Antimicrob Agents Chemother. 54(10):4085–4091.

Tumbarello M, De Pascale G, Trecarichi EM et al. (2013). Clinical outcomes of Pseudomonas aeruginosa pneumonia in intensive care unit patients. Intensive Care Med. 39(4):682–692.

US Centers for Disease Control and Prevention (2013). Antibiotic resistance threats in the United States, 2013. Atlanta, Georgia: US Centers for Disease

Control and Prevention. (http://www.cdc.gov/drugresistance/threat-report-2013, accessed 06 September 2018).

World Health Organization (2014). Antimicrobial resistance: global report on surveillance, 2014. Geneva: World Health Organization. (http://www.who.int/drugresistance/documents/surveillancereport/en/, accessed 06 September 2018).

World Health Organization (2018). Global antimicrobial resistance surveillance system (GLASS) report: early implementation 2016–2017. Geneva: World Health Organization. (http://apps.who.int/iris/bitstream/ handle/10665/259744/9789241513449-eng.pdf;jsessionid=C2EE76D6EE7271D9ED9DF7DDCC771721?sequence=1, accessed 06 September 2018).

3 | Tackling antimicrobial resistance in the community

SARAH TONKIN-CRINE, LUCY ABEL, OLIVER VAN HECKE, KAY WANG, CHRIS BUTLER

Introduction

This chapter provides an overview of why and how antimicrobial resistance (AMR) is being tackled through antimicrobial stewardship (AMS) activities in the community. We discuss the relevance of AMR to antibiotic prescribing in primary care services and for the health professionals who need to engage with AMS activities in order to help tackle AMR. We provide an overview of types of community-level interventions which have been trialled to help promote more prudent use of antibiotics and the evidence behind these. We highlight interventions which currently look to have the most potential and consider how to assess the cost–effectiveness of such interventions. Lastly we assess the challenges to implementing policy on AMS activities at the community-level.

Background

To avoid the increasing burden of AMR, all countries need to implement effective AMS strategies in order to tackle the overuse and misuse of antibiotics. Within the European Union (EU), all antibiotics for systemic use are only available through a prescription written by a qualified health professional. The vast majority of these prescriptions are issued in primary care, rather than secondary or tertiary settings. Across England three quarters of all antibiotics prescribed in 2015 through the National Health Service (NHS) were prescribed for patients seen in a general practice (74%) (Public Health England, 2016). This was followed by hospital inpatients (11%), hospital outpatients (7%), patients seen in dental practices (5%), and patients in other community settings (3%). Therefore, it is important that AMS strategies focus on community settings and target the relevant stakeholders providing and accessing community-based care.

Primary care doctors, or general practitioners (GPs), are the focus of the primary care literature since they are the most frequent prescribers of antibiotics. Nurse practitioners and pharmacists working in community settings also have an important role. In the past 10 years, the role of nurses has expanded to include prescribing in a number of countries and is on the policy agenda in many more (Ball et al., 2009; Hurlock-Chorostecki et al., 2014). Nurse prescribing has been introduced to better utilize the skills and knowledge of health professionals, allow more efficient access to medications and to help reduce the workload of doctors (Courtenay et al., 2014). In the UK, the numbers of nurses qualified to prescribe has steadily increased over the last 5 years and around 31 000 nurses now have the same prescribing capability as doctors (Courtenay et al., 2014). Pharmacists in the UK are also able to register as independent prescribers, usually specializing in prescribing for a particular health condition; for example, diabetes. It is more common for pharmacists to work in secondary care settings, rather than primary. Lastly, dentists are overlooked as prescribers of antibiotics due to the relatively small number of antibiotics prescriptions they give relative to their general practice colleagues. More recently, attention has been paid to dentistry with efforts to promote AMS strategies that encourage more prudent prescribing (Faculty of General Dental Practice, 2016).

Patients presenting in primary care with respiratory, urinary, skin, or tooth infections account for the majority of antibiotic prescriptions. Of these, most antibiotics are prescribed for acute respiratory tract infections (RTIs) (Goossens et al., 2005; Gulliford et al., 2014a; Shapiro et al., 2014). While antibiotics are effective for some RTIs (e.g. community-acquired bacterial pneumonia), the bulk of acute RTIs are self-limiting, as most are of viral origin. Empirical studies have shown that infections such as RTIs and sore throats benefit very little from antibiotics, which often reduce the duration of the symptomatic phase by only a few hours (Smith et al., 2014; Spinks et al., 2013). As such, there is a need to reduce the number of prescriptions for these types of, often viral, infections and empower patients to self-manage their symptoms. For other infections, such as urinary tract infections (UTIs) or skin infections, antibiotics may offer more benefit for patients (Albert et al., 2004; Yue et al., 2016). With these presentations, the aim of AMS strategies may not be to reduce antibiotic prescriptions but rather to encourage narrow-spectrum over broad-spectrum antibiotic use and first-line use where appropriate (Vellinga et al., 2016).

When considering how best to implement AMS, it is important to identify the specific behaviours being carried out by stakeholders in order to target them and encourage change. Health professional behaviour is most often focused on the act of prescribing an antibiotic. Within primary care, this behaviour usually involves a single health professional who assesses the patient and issues the prescription. This is opposed to secondary care, where a team of health professionals may provide a prescription with various actors undertaking different parts of a longer process (Charani et al., 2013). As mentioned before, changing prescribing behaviour can prevent prescription as a whole or involve a change in the prescription type, dose, or duration of treatment.

Once a patient has been given a prescription, they then have to use the prescription, collect the antibiotic, and take the antibiotic. Collecting the prescription and consuming the antibiotic can be seen as two distinct behaviours. However, the latter cannot occur without the former. The (self-reported) consumption of antibiotics is the most common behaviour measured in patients within randomized trials of AMS interventions (Spurling et al., 2017). Alongside antibiotic consumption, it is also important to consider patient behaviour prior to accessing health services. This help-seeking behaviour is potentially more influential on antibiotic prescribing because, if patients do not attend primary care services, they are very unlikely to be able to access antibiotics, ultimately decreasing consumption. Many public campaigns have focused on help-seeking behaviour by the public when implementing AMS strategies (Huttner et al., 2010; Earnshaw et al., 2009; Goossens et al., 2006).

Types of community interventions to tackle AMR and evidence for their effectiveness

Interventions to promote AMS may be identified by the stakeholder groups they target, such as clinicians, patients, or the public. Several multifaceted interventions may target more than one of these groups. The following sections discuss interventions with their main target group(s) in mind when considering the behaviour change of interest. Trials of different interventions are cited as examples of interventions which have worked to change antibiotic prescribing behaviour or consumption behaviour. A description of intervention types, their likely behavioural mechanisms, and evidence for each is presented in Table 3.1.

Table 3.1 *Community behaviour change interventions to target antimicrobial resistance*

Intervention type	Description[a]	Behavioural mechanisms	Example trials or reviews
Clinician-focused interventions			
Clinician education	To include: 1. Educational materials for clinicians: printed, electronic, or audiovisual materials that target the health care professional. 2. Educational meetings: health care professionals attending conferences, lectures, training courses, or workshops. 3. Educational outreach visits: health care professionals receiving information from a trained professional in their practice setting.	Increases clinician knowledge about appropriate antibiotic prescribing. Interactive sessions can increase motivation to change prescribing and increase self-efficacy in prescribing only when indicated.	Van der Velden et al. (2012)
Audit and feedback	Any summary of clinical performance of health care over a specified time period provided to the health care professional.	Allows clinicians to self-monitor prescribing behaviour and to evaluate how well their prescribing matches guidelines and/or their peers. Provides motivation and opportunity to change by highlighting discrepancy in actual and desired behaviour.	Ivers et al. (2012)
Reminders	Verbal, written, or electronic information intended to prompt a health care professional to recall information, to include (computer) decision support systems.	An environmental cue, present at the time of a prescribing decision, designed to interrupt habitual or unconscious processes in clinician prescribing decisions and encourage alternative action.	Garg et al. (2005) Gulliford et al. (2014b)

Financial interventions	Targeting the health care professional (as an individual or a team) to include financial incentives (e.g. fee-for-service) and financial penalties (e.g. direct or indirect financial penalty for inappropriate behaviour).	Increasing clinician motivation to change their prescribing behaviour by incentivizing desired behaviour and/or punishing undesirable behaviour.	Greene et al. (2004) Martens et al. (2006)
Point-of-care tests	Equipment for use by health care professionals in their practice setting, to be used at the time and place of patient care, to provide rapid diagnostic information.	Provides additional clinical information which may decrease clinician uncertainty about diagnosis and/or appropriate management for a specific patient. May also be used as a communication technique to reassure patients that antibiotics are not needed.	Cals et al. (2009) Little et al. (2013) Andreeva & Melbye, 2013)

Clinician- and patient-focused interventions

Enhanced communication training	Any resource targeted at the health care professional and/or patient that encourages discussion about management options to include: 1. clinician-delivered patient educational interventions; 2. improved communication interventions (for clinician–patient interaction); 3. shared decision-making.	Encourages explicit discussion about patient needs and expectations and the benefits and risks of taking antibiotics for the individual in order for the clinician to provide patient-centred care. May increase clinician self-efficacy in discussing management options with patients and may increase patient self-efficacy in self-managing symptoms.	Altiner et al. (2007) Cals et al. (2009) Little et al. (2013) Butler et al. (2012)

Table 3.1 *(cont.)*

Intervention type	Description[a]	Behavioural mechanisms	Example trials or reviews
Patient education materials	Educational materials for patients, or parents of child patients, designed to give new information in printed, electronic, or audiovisual form.	Increases patient or parent knowledge about the illness, symptoms and appropriate management. Likely to provide information about risks and benefits of antibiotics for specific conditions. May increase self-efficacy in self-management of illness.	Francis et al. (2009) Macfarlane Holmes & Macfarlane (1997)
Delayed prescribing strategies	Any resource targeted at the health care professional and/or patient that encourages giving a prescription for a patient to collect or use later than the initial consultation if symptoms do not improve.	Encourages additional explanation from the clinician to increase patient or parent knowledge about the illness and appropriate management. Can increase patient self-efficacy to self-care for their illness and empower patients to decide how to manage their symptoms.	Little et al. (2005) Spiro et al. (2006) De la Poza Abad et al.(2016)
Public-focused interventions			
National Campaigns	Any resource targeted at the health care professional, patient and/or member of the public at the population level employing varied use of communication.	Increases knowledge and awareness of appropriate antibiotic use and antibiotic resistance across several stakeholder groups and may decrease motivation to prescribe or consume antibiotics for self-limiting infections. May increase opportunities for people to discuss the use of antibiotics and/or provide patients with suggesting of questions to ask health care providers.	Huttner et al. (2010) Goossens et al. (2006)

Clinician-focused interventions

There are many types of intervention that have been designed to influence the antibiotic prescribing behaviour of clinicians. Interventions can take the form of a single component (e.g. a guideline) or can be multifaceted, combining a number of components which are complementary (e.g. an intervention utilizing guidelines, reminders, and audit and feedback).

The provision of clinician education is the basis for the majority of interventions. The success of clinical practice guidelines is dependent on their implementation (Carlsen et al., 2007). Guidelines are designed to improve the standard and consistency of health care and assume a knowledge deficit. However, guidelines that improve knowledge alone are unlikely to be enough to encourage significant behaviour change (NICE, 2007). Outreach visits can support guideline implementation by offering clinicians the opportunity to discuss the relevance of guidelines to their own patient population and to learn about the experiences of their peers. Such interaction can increase clinician motivation to change, and increase their confidence in changing their prescribing behaviour, which thereby increases self-efficacy. Research has shown that interventions containing educational meetings can be effective at changing clinician prescribing behaviour (van der Velden et al., 2012).

Audit and feedback involves monitoring clinicians' prescribing practices and then reporting back to the individual about their prescribing patterns. This can be helpful when clinicians underestimate the number or type of prescriptions given and can also enable comparisons between peers to demonstrate how prescribing could be improved safely. Audit and feedback interventions work by increasing motivation to change and by allowing clinicians to self-monitor their own prescribing, providing information which can be used to set clear prescribing goals. A Cochrane review found audit and feedback generally led to small but potentially important improvements in professional practice. However, the effectiveness of audit and feedback appeared to depend on baseline performance and how feedback was provided (Ivers et al., 2012).

Interventions may also involve the use of reminders for clinicians, often incorporated into computer software used within consultations. These systems commonly advise on the recommended treatment for a particular patient based on the information that has been entered. Such reminders can serve as a cue that interrupts habitual behaviour and makes clinicians more conscious of their decision-making process when prescribing. Studies have indicated that computerized decision support systems can improve practitioner performance. Specifically, interventions using such a system have led to decreases in antibiotic prescriptions for RTIs (Garg et al., 2005; Gulliford et al., 2014b; Meeker et al., 2016).

Financial incentives are commonly used to influence clinical practice in areas identified as high priority by health organizations. Previous trials indicate that financial incentives can reduce antibiotic prescribing; however, changes may only be short-term (Greene et al., 2004; Martens et al., 2006). In UK general practice, prudent antibiotic prescribing practices have been endorsed through the introduction of the Quality and Outcomes Framework in 2004 and the Quality Premium in 2015. These initiatives enable general practices to obtain additional funding by meeting pre-set targets, often reducing all antibiotic prescribing by a specific percentage or decreasing the proportion of broad-spectrum antibiotic prescribing. Incentives increase motivation to change behaviour and may also present opportunities to change when supported with other initiatives. The Quality Premium 2015/16 contributed to two million fewer antibiotic prescriptions between April and December 2015 in England, down 7.9% from the previous year (NHS Commissioning Board, 2017).

Interventions may also focus on training clinicians to learn or develop their existing skills to encourage evidence-based prescribing decisions. The use of point-of-care tests (POCTs) in the community aims to provide additional clinical information by which a prescribing decision can be made more easily. These tests can make a clinician more confident in making a diagnosis or prescribing decision which increases their self-efficacy in not providing an antibiotic when it is not indicated. Commonly, trials have focused on testing C-reactive protein (CRP) POCTs and trials have suggested that these are effective in reducing antibiotic prescriptions for RTIs (Cals et al., 2009; Little et al., 2013; Andreeva & Melbye, 2014). Other studies have included the use of procalcitonin and rapid viral diagnostics to help clinicians distinguish between minor and more severe infections, although these have commonly been trialled in emergency departments (Schuetz et al., 2012; Doan et al., 2014). The use of POCTs in community health services varies across Europe, depending on the availability and reimbursement of these tests by health organizations, with tests commonly being used in Scandinavian countries where their costs are reimbursed (Dahler-Eriksen et al., 1997).

Clinician- and patient-focused interventions

Other skill-based interventions have focused on enhanced communication training for clinicians. These interventions also consider the patient role in the consultation and can include intervention components targeted at the patient. Communication training strategies have developed through the understanding that clinicians can overestimate patient expectations for antibiotics, which can contribute to unnecessary prescribing (Butler et al., 1998a; 1998b). Through specific communication techniques, eliciting patient expectations for treatment and concerns about their illness can help a clinician to provide reassurance and information about self-care rather than an unnecessary prescription. Trials testing interventions that contain communication skills training for clinicians, have shown effectiveness in reducing the number of antibiotic prescriptions for the treatment of RTIs (Cals et al., 2009; Little et al., 2013; Altiner et al., 2007; Butler et al., 2012).

Shared decision-making (SDM) is defined as the process of enabling a health professional and patient to make a joint decision about management based on the best available evidence and the patient's values and preferences (Coxeter et al., 2015). SDM, by definition, is specifically designed to target both the clinician and patient. It can involve a variety of techniques including discussing options, communicating benefits and risks, and checking or clarifying understanding (Makoul & Clayman, 2006). SDM is a relatively new term in the literature that also applies to older interventions using the same or similar techniques.

Similar to clinician educational materials, patient educational materials are also used to promote prudent antibiotic prescribing. Such materials may be used within communication-based interventions or alone. Patient educational materials are usually provided at the time of the consultation and may or may not be discussed by the clinician (Francis et al., 2009). In addition, materials may be focused on one type of infection (e.g. sore throats), a patient group (e.g. parents of young children), or on a range of infections across age groups. One large trial testing parent information booklets in UK general practice for children with RTIs showed a reduction in antibiotic prescribing (Francis et al., 2009). However, a previous trial with adult patients presenting with RTIs showed no difference in the trial arm using a patient booklet (Macfarlane, Holmes & Macfarlane, 1997). Patient information booklets have been used as a component of effective communication interventions, which may suggest that patient materials need to be used interactively in the consultation in order to reduce prescribing practices (Little et al., 2013; Francis et al., 2009).

Delayed prescribing (DP) strategies target both clinicians and patients. DP can be implemented in two ways: patients are given the prescription immediately and get specific advice on when to use it or the prescription may be kept "on hold" for the patient to collect after a few days (Little et al., 2005). Interventions promoting DP look to encourage clinicians to issue delayed antibiotic prescriptions rather than immediate antibiotic prescriptions and to change the way the antibiotic prescription is discussed in the consultation. DP is considered appropriate for infections that are associated with self-limiting symptoms (e.g. sore throat, nasal discharge) or infections appearing to be more than a simple viral illness but with no established evidence of a bacterial infection that requires immediate treatment. When given a delayed prescription, a patient is given information about the likely duration of symptoms and is encouraged to only take antibiotics if symptoms continue for longer than expected or if symptoms worsen (Thompson et al., 2013). Following the consultation, DP strategies seek to change patient behaviour by allowing infections to resolve in their own time. This enables patients to learn that symptoms are self-limiting and increases their self-efficacy in management of their symptoms. Trials of DP strategies indicate significantly reduced consumption of antibiotics by patients compared to trial arms providing "immediate prescriptions" (Little et al., 2005; Spiro et al., 2006; De la Poza Abad et al., 2016).

Public-focused interventions

Public-focused interventions most often take the form of national campaigns, which are promoted during winter periods when infections are more prevalent. Most contain messages targeted at the public but may also include components that are tailored to clinicians and specific patient groups. In the UK, campaigns have been running regularly since 1999; however, many more European countries have been encouraged to conduct similar campaigns since the first European Awareness Day on 18 November 2008 (Earnshaw et al., 2009). Evaluations of campaigns in high-income countries, including Belgium and France, suggest that they may help to reduce antibiotic prescribing and consumption, although these studies emphasize that benefits are likely to be seen in countries that are considered high prescribers and only if campaigns use specific behavioural and social marketing techniques to target specific populations (Huttner et al., 2010; Goossens et al., 2006).

Community-level interventions with most potential

As noted above, there have been numerous clinical trials of behaviour change interventions targeted at clinicians, patients, and the public testing their effectiveness in changing antibiotic prescribing or consumption behaviour. To date, certain types of interventions have been trialled more often than others due to initial interest from clinicians and policy-makers, the accessibility and cost of interventions, and the success of previous trials. This section will summarize the evidence to date, for three types of interventions that appear to show promise in tackling AMR in community settings (Table 3.2).

Enhanced communication strategies and shared decision-making

Shared decision-making (SDM) has been identified as a promising approach to tackling AMR since it involves both the clinician and the patient. This strategy can potentially be adapted for any community setting with minimal resources. A Cochrane review of nine randomized controlled trials (RCT) concluded that interventions that facilitated SDM reduced overall antibiotic use (prescription, dispensing, or consumption of antibiotics) for RTI consultations in primary care at time of consultation and up to six weeks after (Coxeter et al., 2015). The authors found that SDM interventions helped reduce antibiotic prescribing without increasing re-consultation for the same illness or affecting patient satisfaction. Seven of the trials were carried out in European general practice and two were carried out in Canadian primary care. Trials included adults and/or children. Another review focused on interventions that reduce antibiotic prescribing for RTIs in children and also identified that interventions which supported clinician–parent interaction in the consultation increased effectiveness in reducing prescribing (Vodicka et al., 2013).

Table 3.2 *A summary of systematic review evidence for three types of community-based antimicrobial stewardship interventions*

Intervention	Review of the evidence	Outcomes	Quality of evidence (GRADE[a])	Summary
CRP point-of-care test versus usual care	Aabenhus, Costa & Vaz-Carneiro (2014)	Change in antibiotic prescription for RTI at consultation: RR 0.78 (0.66 to 0.92)	Moderate	Use of CRP testing probably reduces antibiotic prescribing in general practice and results in little or no difference in patient satisfaction or re-consultation.
		Patient satisfaction: RR 0.79 (0.57 to 1.08)	Moderate	
		Re-consultation: RR 1.08 (0.93 to 1.27)	Moderate	
		Change in antibiotics prescribed or dispensed within 6 weeks of consultation: RR 0.61 (0.55 to 0.68)	Moderate	
Shared decision-making versus usual care	Coxeter et al. (2015)	Patient satisfaction: RR 0.86 (0.57 to 1.30)	Low	Use of shared decision-making probably reduces antibiotic use in general practice and results in little or no difference in patient satisfaction or re-consultation.
		Re-consultation: RR 0.87 (0.74 to 1.03)	Moderate	
		Change in antibiotic use – delayed versus immediate antibiotic prescription: OR 0.04 (0.03 to 0.05)	Moderate	
Delayed prescribing strategies versus immediate prescribing	Spurling et al. (2017)	Patient satisfaction – delayed versus immediate antibiotic prescription: OR 0.65 (0.39 to 1.10)	Moderate	Use of delayed prescriptions probably reduces antibiotic use compared to immediate prescriptions in primary care settings and results in little or no difference in patient satisfaction or re-consultation.
		Re-consultation – delayed versus immediate antibiotic prescription: OR 1.04 (0.55 to 1.98)	Moderate	

[a] GRADE: Grading of Recommendations, Assessment, Development and Evaluation (see https://training.cochrane.org/resource/grade-handbook)

Point-of-care tests

C-reactive protein is the most common POCT that is assessed for its effectiveness in reducing antibiotic prescribing in community settings. Such experiments are typically conducted through control trials with comparison groups. A Cochrane review concluded that CRP testing is an effective way to reduce antibiotic prescribing for RTIs in primary care (Aabenhus, Costa & Vaz-Carneiro, 2014). Studies included in the reviews were carried out most often in European general practices and included adult patients with RTI symptoms. Additional Cochrane reviews have looked at the evidence for the use of procalcitonin and rapid viral diagnostics, indicating the effectiveness of the former but not the latter in decreasing antibiotic prescribing. However, these studies have mainly been conducted in emergency departments (Schuetz et al., 2012; Doan et al., 2014). Recent studies exploring diagnostic POCTs for respiratory viruses indicate that tests could positively influence the prescription of antibiotics by GPs, but that diagnostic accuracy needs to be improved and the influence on clinician decision-making should be further assessed (Bruning et al., 2017). Studies have also explored the implementation of such tests in community pharmacists and identified that offering such a test can improve access to care outside normal clinic hours (Klepser et al., 2017).

Delayed prescribing strategies

In a recent update of a Cochrane review, 11 studies that test DP strategies were identified. The result of DP was compared to both immediate prescribing and no-prescribing strategies for clinical outcomes, antibiotic use, and patient satisfaction (Spurling et al., 2017). Interventions encouraging clinicians to use DP resulted in lower antibiotic use than when an immediate use prescription was given. However, there was no difference between delayed and no-antibiotic prescribing in symptom control or disease complications. Patient satisfaction was greatest when either type of prescription was given. Authors recommended that clinicians should favour no-antibiotic prescribing when they feel confident an antibiotic is not required and encourage patients to re-consult if symptoms do not resolve. However, when clinicians are not confident in using a no-prescribing strategy, DP may help to reduce antibiotic consumption while maintaining patient satisfaction.

Examples of community-level interventions

It is useful to highlight some examples of successful primary care interventions that have been effective in reducing antibiotic prescribing and/or consumption in the community. This section describes three interventions, in detail, to identify the intervention components and mechanisms of behaviour change that contributed to the reduction in antibiotic use.

GRACE INTRO

The Genomics to combat Resistance against Antibiotics for Community-acquired LRTI (lower respiratory tract infection) in Europe/INternet Training for Reducing antibiOtic use (GRACE INTRO) project was an international programme of research carried out across several European countries. A component of GRACE INTRO involved the design and development of a multifaceted intervention to reduce antibiotic prescribing in general practice for acute cough in adults (Little et al., 2013).

The intervention, aimed to train GPs in 1) the use of a CRP POCT during the consultation to inform management decisions and 2) enhanced communication skills, with interactive use of a patient booklet in the consultation, to explain to patients when antibiotics were unlikely to benefit them (Anthierens et al., 2012). CRP training was proposed to help reduce clinician uncertainty about whether a patient would benefit from antibiotics. Clinicians received online tutorials on using the test and interpreting the results, and a visit from a representative of the test manufacturer. A desk reminder was provided to clinicians, giving CRP cut-off values and recommendations for treatment. When an antibiotic is not indicated for patient treatment, communication skills training and interactive use of a booklet was proposed to help clinicians identify patients' needs and concerns which could be addressed with self-management advice and reassurance.

The intervention was tested through a 2×2 factorial RCT across six countries and was shown to be effective at reducing antibiotic prescriptions compared to usual care (Little et al., 2013). Intervention practices received either one or both interventions, with use of both interventions resulting in the greatest decrease in the number of antibiotic prescriptions.

A process evaluation indicated that GPs felt reducing antibiotic prescribing was more important and less risky after taking part in the study. It also found that GPs trained in communication skills were more confident in not prescribing antibiotics for an acute cough (Yardley et al., 2013; Anthierens et al., 2015). Patients in the intervention arms with the interactive booklet reported higher levels of enablement and satisfaction following their consultation compared to other trial arms (Yardley et al., 2013; Tonkin-Crine et al., 2014). Within the CRP intervention arms, there is some evidence that GPs used the tests to convince patients of a no-antibiotic decision rather than as a way to obtain additional clinical information (Tonkin-Crine et al., 2014).

EQUIP

The Enhancing the Quality of Information-sharing in Primary care (EQUIP) project focused on general practice consultations for children with RTIs. The project set out to evaluate whether training clinicians in the use of an interactive parent booklet could influence antibiotic use and rates of re-consultation for the same illness (Francis et al., 2008a).

The intervention used a booklet that was designed for clinicians to discuss with the parents of their patient during a consultation. The booklet went through a vigorous design and development process and included contributions from both parents and GPs. This enhanced its readability and enabled it to meet the required needs of both groups (Francis et al., 2008b). The booklet sought to inform parents about when antibiotics were required and to provide self-care advice for minor infections. It also included safety-netting advice for when parents should consult in primary care. The intervention also included online training for clinicians on how to use the booklet during consultations. This train-ing encouraged clinicians to 1) identify the parents' main concerns and expectations and 2) explicitly discuss prognosis and treatment options. The intervention was tested through a cluster RCT and showed to be effective at reducing antibiotic prescribing by GPs and reducing parents' intention to re-consult without affecting parental satisfaction with care (Francis et al., 2009).

A process evaluation indicated that both clinicians and parents found intervention materials acceptable for use in daily practice (Francis et al., 2013). Intervention materials were thought to increase clinician confidence in discussing a no-prescribing decision and to increase parent confidence in self-caring for their child's RTI. Clinicians reported some barriers to using the booklet interactively in the consultation including lack of familiarity with the booklet, lack of time, and difficulty modifying their consultation style.

Antibiotic Guardian

The UK's "Antibiotic Guardian Campaign", launched in September 2014, aimed to increase awareness and engagement with AMR by health professionals and the public (Ashiru-Oredope & Hopkins, 2015; Chaintarli et al., 2016). The campaign differed from previous UK campaigns in that it was available all year round rather than being seasonal only.

The campaign included a website where people could make online pledges to act to reduce AMR (http://www.antibioticguardian.com). A list of pledges relevant to health professionals or the public was available, and people could choose which pledge they wanted to make. Making an online pledge was hypothesized to bridge the intention–behaviour gap, identified in psychological literature as a barrier to behaviour change. This was accomplished by supporting people when making implementation intentions. These implementation intentions, presented as "if-when plans", help people to identify how they will act in a given situation. Examples for patients include: "If I'm prescribed antibiotics, I will take them exactly as prescribed and never share them with others", and for clinicians: "I will ensure all prescribers in my practice including locums have easy access to the local antibiotic guidance".

The impact of the campaign was assessed via an online survey sent to 9 016 self-selected "Antibiotic Guardians" to assess changes in self-reported knowledge and behaviour (Chaintarli et al., 2016). Two thirds of respondents reported that they had always acted on the pledge they made, around half of participants indicated that their knowledge of AMR had increased due to the campaign, and 70% reported that they felt some personal responsibility for AMR (compared to 58% at baseline).

Results indicated that the Antibiotic Guardian campaigns led to increases in self-reported knowledge of AMR and self-reported behaviour change in line with pledges. A process evaluation of the campaign indicated that people signed up out of personal concern about AMR (Kesten et al., 2018). Pledges encouraged reflection on AMR-related behaviours and keeping to pledges reflected new behaviour change and maintenance of existing behaviours. Responding collectively to a campaign was thought to have a greater impact than individual action. However, respondents felt that the campaign needed greater visibility, especially to engage groups who are less familiar with AMR. Respondents were mostly health care professionals or people who were connected to the health care system and less than a third of respondents pledged as members of the public.

Assessing cost–effectiveness of community interventions

Uptake of AMS interventions in practice relies on a compelling health-economic justification. Health care budgets are limited, so investment in new interventions will inevitably come at the expense of other treatments. Health-economic analysis provides information on how the new intervention compares to what it will replace in terms of costs and benefits, thereby helping health providers align their investment decisions with their overall aims to provide the best possible health outcomes (Drummond, 2005).

There are four components to consider in assessing the cost–effectiveness of AMS strategies. First is effectiveness in reducing antibiotic prescribing. Second is effectiveness in terms of health outcomes. This is important because if a reduction in prescribing results in inferior health outcomes, this will need to be weighed against the value of reducing the health consequences of future AMR as well as against that of alternative interventions that may have improved health outcomes. The third component is cost. Many new interventions, such as POCTs, will cost more than the antibiotics they replace. For example, amoxicillin costs £1.02 for a three-week course while a CRP test costs £5.53, and if additional appointments are required the cost of those extra resources will quickly add up (Joint Formulary Committee, 2018; Hunter, 2015).

Cost–effectiveness studies that assess AMS strategies in terms of the above three components are increasingly common. However, most cost–effectiveness analyses continue to ignore the potential impact on AMR as an outcome or consequence entirely. For example, one study of UTI management evaluated the cost–effectiveness of strategies only in terms of reduction in symptom duration, despite UTI being a strong driver of antibiotic prescribing (Little et al., 2009).

The final component of AMR cost–effectiveness is the value of AMR itself. In economic terms, there is an opportunity cost to preventing AMR in terms of benefits foregone now, such as current health and cost savings. There is considerable uncertainty around both how much society is willing to give up to avoid future AMR, and how much would be necessary to avoid it (Coast et al., 1996). Assuming that not all strategies simultaneously save costs, improve current health, and reduce AMR, these values are required to make a transparent judgement on whether AMS interventions are truly cost-effective.

There have been a small number of studies attempting to consider these outcomes in cost–effectiveness analysis. One study evaluated the proportion of societal costs attributable to AMR from a single prescription of antibiotics, based on global estimates of AMR costs found in three large analyses including the UK AMR review (O'Neill, 2016). They then applied this single cost to each prescription to give some idea of the opportunity cost of antibiotic prescriptions in RTI (Oppong et al., 2016). However, as yet studies are unable to provide valid results on the cost–effectiveness of AMS strategies and considerable methodological work in this area is still required.

Challenges to implementing policy

To date the majority of clinical trials have tested the effectiveness of community interventions which are targeted at general practice settings and focused on reducing antibiotic prescribing for RTIs. The vast majority of these trials have been carried out in high-income countries, with some conducted in middle-income countries such as China (Tonkin-Crine et al., 2017).

Previous trials carried out across Europe have indicated minimal differences in how interventions are accepted and implemented by health professionals and patients (Little et al., 2013). This is encouraging, as interventions have shown to be effective in different health care organizations and in health systems with different financial structures (e.g. services free at the point of care or insurance-based health care). However, the influence of culture and context on antibiotic use is currently underexplored and other studies have highlighted that such factors may be a barrier in transferring effective interventions from one context/country to another (Touboul-Lundgren et al., 2015). The current evidence in this area is limited in how readily it can apply to other low- and middle-income countries (Tonkin-Crine et al., 2017). Interpreting evidence for these settings is a barrier to policy-makers as there is a limited understanding of the contextual factors that influence antibiotic prescribing behaviour and antibiotic consumption behaviour.

Policy-makers should be cautious about assuming that an effective intervention in one context will be effective in another given differences in health care organization, culture, and/or country.

The evidence base is also limited for the long-term impact of interventions. Many trials have focused on short-term outcomes, either a few weeks or months post-intervention. Although these results are positive, it is difficult to establish whether interventions lead to long-term behaviour change or whether clinicians and patients eventually return to habitual pre-trial behaviours. Larger trials have observed interventions applied in clinical practice from 1 to 3 years to explore the subsequent long-term effect on prescribing rates (Little et al., 2005; Cals et al., 2013). These trials suggest that particular types of intervention are potentially more likely to support long-term behaviour change than other types. For example, the use of enhanced communication strategies is more likely to have an effect long-term than the use of CRP tests when reducing antibiotic prescribing for RTIs (Little et al., 2005; Cals et al., 2013). This suggests that interventions based on enhancing the skills of health professionals may be implemented more easily than use of novel technologies as there is potentially less disruption to clinical practice, and skills can be rehearsed and learnt more easily. The impact of the long-term effects of interventions needs to be researched more thoroughly and again may differ depending on the context of interest.

Conclusions

Interventions aimed at tackling AMR can target a number of behaviours carried out by different stakeholders, including in the course of consulting, prescribing, dispensing and consumption of antibiotics. Policy-makers wanting to tackle AMR should identify the specific behaviours that are going to have the greatest impact. To date, the literature has focused on RTIs in general practice, which account for the vast majority of antibiotic prescribing in Europe. However, for different contexts and countries, the target behaviour may be very different.

There are a number of influences on antibiotic prescribing and consumption behaviours. The clinical factors at patient presentation can be very similar between contexts; however, the social, cultural and environmental factors may be significantly different. Interventions need to address all of these influences to be effective at changing behaviour. As such, interventions being trialled in new contexts must take into account the cultural and social preferences of the groups whose behaviour they are trying to change.

Community interventions that tackle AMR require further testing in primary care contexts outside general practice and in low- and middle-income countries where little is known about the influences on antibiotic-related behaviours.

References

Aabenhus R, Jensen JU, Jørgensen KJ et al. (2014). Biomarkers as point-of-care tests to guide prescription of antibiotics in patients with acute respiratory infections in primary care. Cochrane Database Syst Rev. 11:CD010130.

Albert X, Huertas I, Pereiró II et al. (2004). Antibiotics for preventing recurrent urinary tract infection in non-pregnant women. Cochrane Database Syst Rev. 4:CD001209.

Altiner A, Brockmann S, Sielk et al. (2007). Reducing antibiotic prescriptions for acute cough by motivating GPs to change their attitudes to communication and empowering patients: a cluster-randomized intervention study. J Antimicrob Chemother. 60(3):638–644.

Andreeva E, Melbye H (2014). Usefulness of C-reactive protein testing in acute cough/respiratory tract infection: an open cluster-randomised clinical trial with CRP testing in the intervention group. BMC Family Pract. 15:80.

Anthierens S, Tonkin-Crine S, Douglas E et al. (2012). General practitioners' views on the acceptability and applicability of a web-based intervention to reduce antibiotic prescribing for acute cough in multiple European countries: a qualitative study prior to a randomised trial. BMC Family Pract. 13:101.

Anthierens S, Tonkin-Crine S, Cals JW et al. (2015). Clinicians' views and experiences of interventions to enhance the quality of antibiotic prescribing for acute respiratory tract infections. J Gen Int Med. 30(4):408–416.

Ashiru-Oredope D, Hopkins S (2015). Antimicrobial resistance: moving from professional engagement to public action. J Antimicrob Chemother. 70(11):2927–2930.

Ball J, Barker G, Buchanan J (2009). Implementing nurse prescribing: an updated review of current practice internationally. Geneva: International Council of Nurses.

Bruning AH, de Kruijf WB, van Weert HCPM et al. (2017). Diagnostic performance and clinical feasibility of a point-of-care test for respiratory viral infections in primary health care. Fam Pract. 34(5):558–563.

Butler CC, Rollnick S, Kinnersley P et al. (1998a). Reducing antibiotics for respiratory tract symptoms in primary care: consolidating "why" and considering "how". Br J Gen Pract. 48(437):1865–1870.

Butler CC, Rollnick S, Pill R et al. (1998b). Understanding the culture of prescribing: qualitative study of general practitioners' and patients' perceptions of antibiotics for sore throats. BMJ. 317:637–642.

Butler CC, Simpson SA, Dunstan F et al. (2012). Effectiveness of multifaceted educational programme to reduce antibiotic dispensing in primary care: practice based randomised controlled trial. BMJ. 344:d8173.

Cals JW, Butler CC, Hopstaken RM et al. (2009). Effect of point of care testing for C reactive protein and training in communication skills on antibiotic use in lower respiratory tract infections: cluster randomised trial. BMJ. 338:b1374.

Cals JWL, de Bock L, Beckers PJ et al. (2013). Enhanced communication skills and C-reactive protein point-of-care testing for respiratory tract infection: 3.5-year follow-up of a cluster randomized trial. Ann Fam Med. 11(2):157–164.

Carlsen B, Glenton C, Pope C (2007). "Thou shalt versus thou shalt not": a meta-synthesis of GPs' attitudes to clinical practice guidelines. Br J Gen Pract. 57(545):971–978.

Chaintarli K, Ingle SM, Bhattacharya A et al. (2016). Impact of a United Kingdom-wide campaign to tackle antimicrobial resistance on self-reported knowledge and behaviour change. BMC Pub Health. 16:393.

Charani E, Castro-Sanchez E, Sevdalis N et al. (2013). Understanding the determinants of antimicrobial prescribing within hospitals: The role of "prescribing etiquette". Clin Infect Dis. 57(2):188–196.

Coast J, Smith RD, Millar MR (1996). Superbugs: Should antimicrobial resistance be included as a cost in economic evaluation? Health Econom. 5(3):217–226.

Courtenay M, Gillespie D, Lim R (2017). Patterns of dispensed non-medical prescriber prescriptions for antibiotics in primary care across England: a retrospective analysis. J Antimicrob Chemother. 72(10):2915–2920.

Coxeter P, Del Mar CB, McGregor L et al. (2015). Interventions to facilitate shared decision making to address antibiotic use for acute respiratory infections in primary care. Cochrane Database Syst Rev. 11:CD010907.

Dahler-Eriksen BS, Lassen JF, Lund ED et al. (1997). C-reactive protein in general practice – how commonly is it used and why? Scand J Prim Health Care. 15(1):35–38.

De la Poza Abad M, Mas Dalmau G, Moreno Bakedano M et al. (2016). Prescription strategies in acute uncomplicated respiratory infections. JAMA. 176(1):21–29.

Doan Q, Enarson P, Kissoon N et al. (2014). Rapid viral diagnosis for acute febrile respiratory illness in children in the Emergency Department. Cochrane Database Syst Rev. 9:CD006452.

Drummond MF (2005). Methods for the economic evaluation of health care programmes, 3rd edn. Oxford: Oxford University Press.

Earnshaw S, Monnet DL, Duncan B et al. (2009). European Antibiotic Awareness Day, 2008 – the first Europe-wide public information campaign on prudent antibiotic use: methods and survey of activities in participating countries. Euro Surveill. 14(30):19280.

Faculty of General Dental Practice (2016). Antimicrobial prescribing for GDPs. London: Faculty of General Dental Practice. (https://www.fgdp .org.uk/guidance-standards/antimicrobial-prescribing-gdps accessed 05 December 2019).

Francis NA, Hood K, Simpson S et al. (2008a). The effect of using an interactive booklet on childhood respiratory tract infections in consultations: Study protocol for a cluster randomised controlled trial in primary care. BMC Family Pract. 9(1):23.

Francis NA, Wood F, Simpson S et al. (2008b). Developing an "interactive" booklet on respiratory tract infections in children for use in primary care consultations. Patient Educ Couns. 73(2):286–293.

Francis NA, Butler CC, Hood K et al. (2009). Effect of using an interactive booklet about childhood respiratory tract infections in primary care consultations on re-consulting and antibiotic prescribing: a cluster randomised controlled trial. BMJ. 339:374–376.

Francis NA, Phillips R, Wood F et al. (2013). Parents' and clinicians' views of an interactive booklet about respiratory tract infections in children: a qualitative process evaluation of the EQUIP randomised controlled trial. BMC Family Pract. 14:182.

Garg AX, Adhikari NK, McDonald H et al. (2005). Effects of computerized clinical decision support systems on practitioner performance and patient outcomes: a systematic review. JAMA. 293:1223–1238.

Goossens H, Ferech M, Vander Stichele R et al. (2005). Outpatient antibiotic use in Europe and association with resistance: a cross-national database study. Lancet. 365(9459):579–587.

Goossens H, Guillemot D, Ferech M et al. (2006). National campaigns to improve antibiotic use. Eur J Clin Pharmacol. 62(5):373–379.

Greene RA, Beckman H, Chamberlain J et al. (2004). Increasing adherence to a community-based guideline for acute sinusitis through education, physician profiling, and financial incentives. Am J Manag Care. 10:670–678.

Gulliford MC, Dregan A, Moore MV et al. (2014a). Continued high rates of antibiotic prescribing to adults with respiratory tract infection: survey of 568 UK general practices. BMJ Open 4:e006245.

Gulliford MC, van Staa T, Dregan A et al. (2014b). Electronic health records for intervention research: A cluster randomized trial to reduce antibiotic prescribing in primary care (eCRT Study). Ann Fam Med. 12(4):344–351.

Hunter R (2015). Cost-effectiveness of point-of-care C-reactive protein tests for respiratory tract infection in primary care in England. Adv Ther. 32 (1):69–85.

Hurlock-Chorostecki C, Forchuk C, Orchard C et al. (2014). Labour saver or building a cohesive interprofessional team? The role of the nurse practitioner within hospitals. J Interprof Care. 28:260–266.

Huttner B, Goossens H, Verheij T et al. (2010). Characteristics and outcomes of public campaigns aimed at improving the use of antibiotics in outpatients in high-income countries. Lancet Infect Dis. 10(1): 17–31.

Ivers N, Jamtvedt G, Flottorp S et al. (2012). Audit and feedback: effects on professional practice and healthcare outcomes. Cochrane Database Syst Rev. 6:CD000259.

Joint Formulary Committee (2018). British National Formulary (online). London: BMJ Group and Pharmaceutical Press. (http://www .medicinescomplete.com, accessed 06 September 2018).

Kesten JM, Bhattacharya A, Ashiru-Oredope D et al. (2018). The Antibiotic Guardian campaign: a qualitative evaluation of an online pledge-based system focused on making better use of antibiotics. BMC Pub Health. 18:5.

Klepser DG, Klepser ME, Smith JK et al. (2017). Utilization of influenza and streptococcal pharyngitis point-of-care testing in the community pharmacy practice setting. Res Social Adm Pharm. 14(4):356–359.

Little P, Rumsby K, Kelly J et al. (2005). Information leaflet and antibiotic prescribing strategies for acute lower respiratory infection. JAMA. 293(24):3029–3035.

Little P, Turner S, Rumsby K et al. (2009). Dipsticks and diagnostic algorithms in urinary tract infection: development and validation, randomised trial, economic analysis, observational cohort and qualitative study. Health Tech Assess. 13(19):1–73.

Little P, Stuart B, Francis N et al. (2013). Effects of internet-based training on antibiotic prescribing rates for acute respiratory-tract infections: a multinational, cluster, randomised, factorial, controlled trial. Lancet. 382(9899):1175–1182.

Little P, Stuart B, Francis N (2017). Antibiotic prescribing for acute respiratory tract infections 12 months after internet-based training in communication skills and an interactive patient booklet, and in the use of a CRP point-of-care test: a multi-national cluster-randomised controlled trial. Personal communication.

Macfarlane JT, Holmes WF, Macfarlane RM (1997). Reducing reconsultations for acute lower respiratory tract illness with an information leaflet: a randomized controlled study of patients in primary care. Br J Gen Pract. 47:719–722.

Makoul G, Clayman ML (2006). An integrative model of shared decision making in medical encounters. Patient Educ Couns. 60:301–312.

Martens JD, Werkhoven MJ, Severens JL et al. (2006). Effects of a behaviour independent financial incentive on prescribing behaviour of general practitioners. J Eval Clin Pract. 13:369–373.

Meeker D, Linder JA, Fox CR et al. (2016). Effect of behavioral interventions on inappropriate antibiotic prescribing among primary care practices: A randomized clinical trial. JAMA. 315(6):562–570.

National Institute for Health and Care Excellence (NICE) (2007). Behaviour change: general approaches. NICE public health guidance 6. London: NICE. (https://www.nice.org.uk/guidance/PH6, accessed 06 September 2018).

NHS Commissioning Board (2017). Antibiotic prescribing quality premium 2016/17. London: NHS Commissioning Board. (http://medicines.necsu.nhs.uk/antibiotic-prescribing-quality-premium-201617/, accessed 06 September 2018).

O'Neill J (2016). Tackling drug-resistant infections globally: final report and recommendations. The Review on Antimicrobial Resistance. London: Wellcome Trust and Government of the United Kingdom. (https://amr-review.org/Publications.html, accessed 06 September 2018).

Oppong R, Smith RD, Little P et al. (2016). Cost effectiveness of amoxicillin for lower respiratory tract infections in primary care: an economic evaluation accounting for the cost of antimicrobial resistance. Br J Gen Pract. 66(650):e633–e639.

Public Health England (PHE) (2016). English Surveillance Programme for Antimicrobial Utilisation and Resistance (ESPAUR) 2010–2015: report 2016. London: Public Health England. (https://assets.publishing.service.gov.uk/government/uploads/system/uploads/attachment_data/file/575626/ESPAUR_Report_2016.pdf, accessed 06 September 2018).

Schuetz P, Müller B, Christ-Crain M et al. (2012). Procalcitonin to initiate or discontinue antibiotics in acute respiratory tract infections. Cochrane Database Syst Rev. 9:CD007498.

Shapiro DJ, Hicks LA, Pavia AT et al. (2014). Antibiotic prescribing for adults in ambulatory care in the USA, 2007–09. J Antimicrob Chemother. 69(1):234–240.

Smith SM, Fahey T, Smucny J et al. (2014). Antibiotics for acute bronchitis. Cochrane Database Syst Rev. 3:CD000245.

Spinks A, Galsziou PP, Del Mar CB (2013). Antibiotics for sore throat. Cochrane Database Syst Rev. 11:CD000023.

Spiro DM, Tay KY, Arnold DH et al. (2006). Wait-and-see prescription for the treatment of acute otitis media: a randomized controlled trial. JAMA. 296(10):1235.

Spurling GKP, Del Mar CB, Dooley L et al. (2017). Delayed antibiotic prescriptions for respiratory infections. Cochrane Database Syst Rev. 9:CD004417.

Thompson M, Vodicka TA, Blair PS et al. (2013). Duration of symptoms of respiratory tract infections in children: systematic review. BMJ. 347:f7027.

Tonkin-Crine S, Anthierens S, Francis NA et al. (2014). Exploring patients' views of primary care consultations with contrasting interventions for acute cough: a six-country European qualitative study. Prim Care Resp Med. 24:14026.

Tonkin-Crine S, Anthierens S, Hood K et al. (2016). Discrepancies between qualitative and quantitative evaluation of randomised controlled trial results: achieving clarity through mixed methods triangulation. Implement Sci.11:66.

Tonkin-Crine S, Tan PS, van Hecke O et al. (2017). Clinician-targeted interventions to influence antibiotic prescribing behaviour for acute respiratory infections in primary care: an overview of systematic reviews. Cochrane Database Syst Rev. 9:CD012252.

Touboul-Lundgren P, Jensen S, Drai J et al. (2015). Identification of cultural determinants of antibiotic use cited in primary care in Europe: A mixed research synthesis study of integrated design "culture is all around us". Health behavior, health promotion and society. BMC Pub Health. 15(1):908.

van der Velden AW, Pijpers EJ, Kuyvenhoven MM et al. (2012). Effectiveness of physician-targeted interventions to improve antibiotic use for respiratory tract infections. Br J Gen Pract. 62(605):e801–e807.

Vellinga A, Galvin S, Duane S et al. (2016). Intervention to improve the quality of antimicrobial prescribing for urinary tract infection: a cluster randomized trial. CMAJ. 188(2):108–115.

Vodicka TA, Thompson M, Lucas P et al. (2013). Reducing antibiotic prescribing for children with respiratory tract infections in primary care: a systematic review. Br J Gen Pract. 63(612):e445– e454.

Yardley L, Douglas E, Anthierens S et al. (2013). Evaluation of a web-based intervention to reduce antibiotic prescribing in six European countries: quantitative process analysis of the GRACE/INTRO randomised controlled trial. Implement Sci. 8:134.

Yue J, Dong BR, Yang M et al. (2016). Linezolid versus vancomycin for skin and soft tissue infections. Cochrane Database Syst Rev. 1:CD008056.

4 | Tackling antimicrobial resistance in the hospital sector

RASMUS LEISTNER, INGE GYSSENS

Introduction

Antibiotic use is a major driving force behind antimicrobial resistance (AMR). Inappropriate use and poor infection prevention and control (IPC) are fuelling the increased resistance. The importance of AMR within the hospital sector is considerable because of the high volumes of antimicrobial substances used by relatively small populations.

Surveillance programmes are a crucial component of antibiotic stewardship for benchmarking antibiotic consumption and detecting possible outbreaks of resistance. Notification of outbreaks with resistant bacteria can also improve the effectiveness of EU early warning systems. In Europe, the European Centre for Disease Prevention and Control (ECDC) runs two surveillance systems, one on antibiotic consumption (European Surveillance of Antimicrobial Consumption Network (ESAC-Net)) and one on antibiotic resistance (European Antimicrobial Resistance Surveillance Network (EARS-Net)). EARS-Net is the largest publicly funded AMR surveillance system in Europe and was established in 1998 by the European Commission (Gagliotti et al., 2011). Its data are based on routine laboratory data from many participating European countries (de Kraker & van de Sande-Bruinsma, 2007). The laboratories report the results from microbiological diagnostics and susceptibility testing of blood cultures and cerebrospinal fluid. Because the data are collected and analysed continuously over time, it can reveal potential trends in AMR across Europe. However, this microbiological surveillance is limited as it lacks epidemiological, clinical, or outcome data (Tacconelli et al., 2017).

Reliable estimates of excess morbidity, mortality, and the costs of AMR must be put into perspective against other causes. Apart from malaria, tuberculosis, gonorrhoea, and HIV, most of the disease burden attributable to AMR is caused by health care-associated infections (HAIs) due to opportunistic bacteria. In the Burden of Resistance and Disease in European Nations (BURDEN) project, de Kraker et al. estimated the impact on AMR of the two most frequent causes of bloodstream

71

infections (BSIs) worldwide – *Staphylococcus aureus* and *Escherichia coli* – in 13 European hospitals (de Kraker et al., 2011a; de Kraker et al., 2011b). These data were extrapolated to 31 countries that participated in the European Antibiotic Resistance Surveillance System (EARSS). It was estimated that in 2007 over 8 000 deaths and €62 million in excess costs were associated with BSIs caused by methicillin-resistant *S. aureus* (MRSA) and *E. coli* resistant to third-generation cephalosporins (G3REC) in the European Region. For G3REC and MRSA BSIs in the high-income Organisation for Economic Co-operation and Development (OECD) countries, the estimated mortality of 1.5 per 100 000 is comparable with rates for HIV/AIDS (1.5 per 100 000) or tuberculosis (1.0 per 100 000). The authors conclude that mortality attributed to AMR is high, but not excessive when compared to other conditions. The prolongation of hospital stays imposes a considerable burden on health care systems (de Kraker et al., 2011c).

The stakeholders of relevance to AMR in hospitals are the prescribing doctors, pharmacists, nurses, infection control practitioners, managers and members of the hospital board. Leadership support is critical to the success of antibiotic stewardship programmes (Fridkin & Srinivasan, 2014). All stakeholders need to make efforts to implement appropriate antibiotic management and infection prevention to curb AMR and its spread.

This chapter reviews the two main pillars of good practice for mitigation and control of AMR: infection prevention and control and antibiotic stewardship (ABS). To illustrate certain concepts, the analysis focuses on OECD countries as case examples.

Infection control

Approximately 6% of European patients develop a HAI (ECDC, 2013). Lower respiratory tract infections, urinary tract infections (UTIs), surgical site infections, and bloodstream infections account for 75% of HAIs. A number of pathogens have tested positive for resistance to clinically important antibiotic substances. For example, 41% of *S. aureus* are methicillin-resistant and 33% of Enterobacteriaceae are third-generation cephalosporin-resistant. Medically effective measures, which are also cost-effective, are necessary to reduce the number of HAIs and prevent resistance from spreading within hospitals.

Infection prevention and control should be organized centrally at the hospital level using a dedicated team of nurses and physicians, microbiological support, and data analysis support (Zingg et al., 2015). National guidelines, along with continuous education and training, provide up to date standards of care for prevention and control of HAIs throughout hospitals. Both interventions have been associated with lower HAI infection rates following implementation. More positive attitudes are generally found among nurses in paediatric intensive-care units (ICUs) than among physicians in adult ICUs. The uptake of such measures by health care professionals is most successful if they are part of a multimodal intervention, simulation-based training, or hands-on training workshops.

Infection prevention measures can be horizontal or vertical. Horizontal measures are general measures affecting an entire institution; for example, the implementation of a multimodal approach to improved hand hygiene (Pittet et al., 2000). Vertical measures tackle specific problems, such as a policy to reduce central venous catheter-associated bloodstream infections (Huang et al., 2013). In addition to both horizontal and vertical measures, it has been further shown that participation in a prospective surveillance system, regular feedback, and networking can lead to an impressive decline in HAI rates (Zingg et al., 2015). This success has been seen with the German KISS, the Dutch PREZIES, and the French ReAct surveillance systems.

Infection control measures

Several measures have been used to reduce the prevalence of HAIs and antibiotic-resistant bacteria (ARBs) in hospitals:

- *Standard and contact precautions*: Standard precautions are applied in hospitals in order to prevent basic infection. This includes hand hygiene policies and the use of personal protective equipment. Contact precautions are used in addition to standard precautions, comprising measures aimed at the discontinuation of pathogen-specific transmission pathways. These measures can include gowning, gloving, wearing a mask, and using patient-dedicated non-critical care equipment (e.g. stethoscopes) (Tacconelli et al., 2014).
- *Isolation or single-room care*: If a patient is infected or colonized with the targeted pathogen, the patient can be transferred to a single room or into a cohort isolation together with patients colonized by the same pathogen. An alert code for patients previously colonized with ARB following single-room isolation has proven to be an effective strategy in preventing further spread of ARBs. This pre-emptive isolation remains active until the current colonization status of the patient has been verified. However, in the context of increasing

colonization rates, there is an ongoing discussion on whether or not contact precautions are both effective and cost-effective, especially concerning Gram-negative bacteria (Tschudin-Sutter et al., 2017).

- *Active screening cultures*: Since many ARBs spread within the community, the number of patients with undetected colonization status can be highly dependent on the pathogen, setting, and country (Harris et al., 2004). In order to prevent hospital-wide spread of ARBs, active screening for colonization followed by strict contact precautions are recommended (Weintrob et al., 2010). For Gram-negative pathogens, the combination of screening the perirectal and groin areas results in the detection of 95% of carriers. For MRSA, the combination of screening throat and groin areas detects approximately 90% of carriers (Marshall & Spelman, 2007). This evidence indicates the need for active screening procedures as a prevention strategy since the majority of patients that enter the hospital test positive for ARBs.

- *Environmental cleaning (EC)*: Cleaning of particular surfaces near infected or colonized patients has been shown to be fundamental in HAI prevention and control (Barker, Alagoz & Safdar, 2017; Dancer, 2011). However, the pathogens on dry hospital surfaces vary in their resilience to EC. Strong evidence for EC effectiveness has been demonstrated by the control of outbreaks of *Acinetobacter baumannii* (Tankovic et al., 1994). Other examples of EC strategies have been used for *Pseudomonas aeruginosa*, which is spread via various pathways but most typically originates in biofilms in sinks (Salm et al., 2016). However, there is minimal evidence proving the effectiveness of EC in preventing HAIs in an endemic setting. EC is primarily used in a bundled approach and it is therefore difficult to assess its effectiveness as a single measure (Tacconelli et al., 2014).

- *Universal decolonization*: This strategy is used for reducing the rates of HAIs. It has shown to be successful in preventing bloodstream infections, such as extended spectrum beta-lactamase-positive Enterobacteriaceae (Huang et al., 2013). In this approach, all patients, regardless of their colonization status, receive a daily chlorhexidine bath and mupirocin nasal ointment. Apart from this, there are no other promising regimens for long-term eradication of Gram-negative gut pathogens (Tacconelli et al., 2014).

Cost–effectiveness of infection control measures to prevent HAIs

A 2007 study evaluated the complex relationship between the rate of HAIs and the cost–effectiveness of preventive measures (Halton & Graves,

2007). The study model presupposed the effectiveness of the prevention measure under evaluation. The findings demonstrated that infections due to ARBs led to a prolonged length of hospital stay, associated with increased costs for additional diagnostics, therapeutic interventions, and the additional number of hospital bed days. These effects derive from both patient complications and blocking of beds to prevent further patient contact and infection. These costs have a greater impact on hospitals that operate on a diagnosis-related group (DRG)-based system since prospective remuneration on admission is not typically covered by insurance companies. In order to assess the cost–effectiveness of a prevention measure, the excess costs of the HAI under consideration and the necessary investment to prevent the infection need to be known (Figure 4.1).

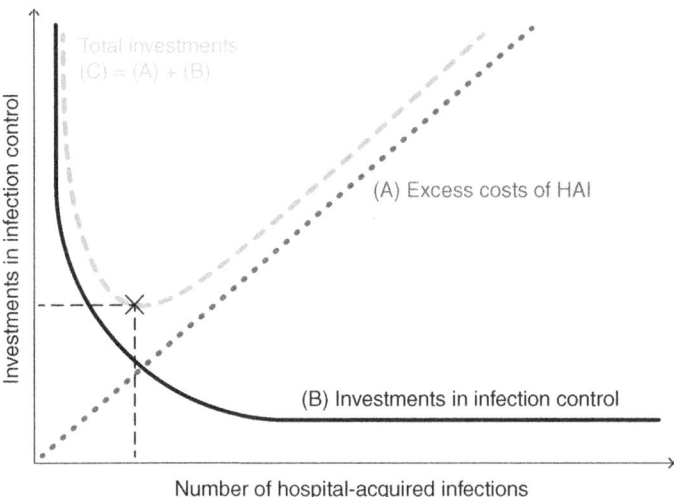

Figure 4.1 Relationship between the number of hospital-acquired infections and investments in infection control

Notes: HAI: hospital-acquired infection.

Line A (dotted) represents the costs of hospital infections, which is also the savings that would result from prevention. Line B (solid line) summarizes the relationship between the cost and the effectiveness of infection control strategies. Line C (dashes) is the sum of lines A and B for every incidence rate of hospital infections, representing the total cost of infection control strategies for HAIs. The point "X" represents the incidence of infection that minimizes total cost, which indicates a rational objective for decision-makers.

Source: Graves, Plowman & Roberts, 2001.

Robust data on the excess costs associated with HAIs are scarce. Currently, there is one meta-analysis that has estimated the excess costs of HAIs in the USA between 1986 and 2013 (Zimlichman et al., 2013). This study found that the additional costs range from approximately $900 (catheter-related UTI) to $46 000 (central line-associated bloodstream infection due to MRSA) (Table 4.1). These data should be interpreted cautiously as it represents the costs in only one national health care system. However, since most OECD countries use DRG-based payment systems, and in the absence of more reliable data, these figures can serve as orientation (OECD, 2014).

The most expensive aspects of a DRG-based hospital payment system are blocked beds and prolonged stays due to complications from HAIs. Beds are blocked in a multi-bed room in order to isolate an infected patient, leading to the non-availability of the remaining beds and reduced occupancy rates. Together, both infection control strategies account for approximately 80% of excess costs related to AMR (Conterno et al., 2007; Hübner et al., 2014). Although HAIs are commonly associated with additional diagnostics and treatment, the main driver of excess financial costs is the daily cost of hospital beds (Shepard et al., 2013).

Table 4.1 *Costs and length of stay in days by health care-associated infection type*

Health care-associated infection type	Cost ($)	LOS (days)
Surgical site infections	20 785	11.2
MRSA	42 300	23.0
Central line-associated bloodstream infections	45 814	6.9–10.4
MRSA	58 614	15.7
Catheter-associated UTIs	896	not relevant
Ventilator-associated pneumonia	40 144	8.4–13.1
Clostridium difficile infections	11 285	3.3

Notes: Data are reported as means.

LOS: Length of stay; UTI: urinary tract infection.

Source: Zimlichman et al., 2013.

Interventions to prevent cross transmission

Similar to other bacteria, ARBs are transmitted within the hospital predominantly via patient contact with the hands of their caregivers (Longtin et al., 2011; Pittet et al., 2000; Pittet et al., 2006; Tacconelli et al., 2014). Patients who are infected or colonized with ARBs carry billions of colony-forming units per millilitre of stool or sputum. Hand hygiene involves cleaning hands with an alcohol-based hand rub to prevent the spread of bacteria. Health care workers who do not rigorously do this can carry hundreds of thousands of colony-forming units on their hands, which can then be transmitted to other patients. In the case of ARBs, this leaves the affected patients prone to diminished treatment options if exogenous infection occurs (Sax et al., 2007). Although this mode of pathogenesis is well-known and accepted in the medical world, compliance with hand hygiene policies by health care workers in hospitals is often as low as 40% (Longtin et al., 2011). In order to facilitate the promotion of good hand hygiene in hospitals, the WHO developed an educational tool consisting of five indications of when hands should be disinfected: before patient contact, before an aseptic task, after exposure to bodily fluids, after patient contact, and after contact with patient surroundings (Sax et al., 2007). However, to improve compliance, multimodal strategies or intervention bundles should be used as they are found to be more successful (Damschroder et al., 2009). Most importantly, a positive organizational culture is connected to low HAI rates and stabilizing high levels of hand hygiene compliance. Although difficult to assess, the success of this type of culture seems to be associated with the existence of role models who engage in hand hygiene and infection prevention.

From the late 1980s to early 2000s, health care systems in many industrialized countries have been restructured with the goal of decreasing hospital costs and increasing productivity (Clements et al., 2008). This most often leads to shorter hospital stays per patient, enhanced patient throughput and hospital capacity. At the same time, AMR has been on the rise for MRSA, extended spectrum beta-lactamases (ESBL), vancomycin-resistant enterococci, and carbapenem-resistant Enterobacteriaceae (ECDC, 2010; Gagliotti et al., 2011; Gastmeier et al., 2014; de Kraker & van de Sande-Bruinsma, 2007). This has created a vicious cycle characterized by overcrowding and understaffing that works against high levels of hand hygiene compliance. This eventually led to increased HAI rates, hospital costs, and more cost pressure (Clements et al., 2008). Although many countries have acknowledged this as a problematic situation, few have taken action or initiated measures to relieve the pressure (Kaier, Mutters & Frank, 2012).

Surveillance systems of HAIs

In many industrialized countries, HAIs are benchmarked to allow for comparison across different hospitals (Haustein et al., 2011; Tacconelli et al., 2017). This has the potential to identify best practices, improve standards of care, and stabilize the performance of the health services offered. A useful comparison of outcome indicators requires consistent definitions, surveillance methods, and standardized rates. These rates must also be risk-adjusted for differences across the patient population and types of medical procedures. In Europe, there are several national surveillance systems for HAIs and a centralized surveillance system for ARBs – EARS-Net. In 2011–2012 and in 2016–2017, the ECDC also carried out two Europe-wide point prevalence surveys (PPS) of HAIs and antimicrobial use in acute hospitals (ECDC, 2013; 2016).

Surveillance definitions for HAIs are somewhat complex and can lead to disagreement among the clinicians responsible; for example, with ventilator-associated pneumonia and surgical site infection. Surveillance systems should therefore be the responsibility of professionals trained in HAI surveillance (Gastmeier et al., 2006). Continuous surveillance systems that assess primarily for infection incidence are time-consuming and often cost-effective only for larger institutions. By contrast, measuring infection prevalence with cross-sectional surveys (e.g. PPS) is less resource-intensive (Haustein et al., 2011; Tacconelli et al., 2017).

However, this system is more applicable for assessing the overall burden of HAIs than for benchmarking between hospitals. The use of HAI indicators for benchmarking in the different national surveillance systems is well established, yet there are substantial differences with respect to the indicators, methods, and reporting techniques. An example of a successfully functioning national surveillance system is the German KISS system (Krankenhaus-Infektions-Surveillance-System). Based on the US national nosocomial infections surveillance system model, this voluntary and confidential system has been in operation since 1997 (Gastmeier et al., 2008). Currently, approximately two thirds of all

German hospitals participate in KISS (Schröder et al., 2015; Leistner et al., 2013). This system provides several small surveillance modules for various risk groups (e.g. ICU patients, surgical patients), ARBs, and C. *difficile*, and the use of an alcohol-based hand rub.

ECDC collects laboratory-based data for AMR using the European Antimicrobial Resistance Surveillance Network (EARS-Net). Although all existing national and international surveillance systems can provide comprehensive information on AMR and HAIs, the results are usually published years after the data are collected (Tacconelli et al., 2017). This diminishes their utility for clinical, institutional, and regulatory decision-making. Further, the delay may result in misalignment of targeted resources and research priorities since information will not be up to date at the time of policy development.

Measures for outbreak control

Outbreaks of HAIs pose major challenges for hospital management and the infection control department of an affected institution. Although outbreaks typically affect only one department or ward, they are often publicly perceived as a malfunction of the entire hospital. In order to control current and future outbreaks, such events should be rigorously investigated. Infection and microbiological diagnoses should be recorded and analysed continuously. This can act as an early warning system. Such systems can be based on microbiology data from the laboratory, surveillance data, or clinical data. The delay between a possible outbreak and its detection depends on the type of system used. In order to ensure cost–effectiveness of the system, it should be adapted to the individual hospital based on a risk assessment by an infection control or hospital epidemiology expert.

The results of the analysis should then be communicated to all affected players since this information presents an excellent training opportunity. In the case of an outbreak, an alert signal is provided by the surveillance system. The outbreak alert can result from clusters of:

- the same pathogen (e.g. ARBs in different microbiological specimens) (Salm et al., 2016);
- the same types of infection (e.g. central venous catheter-associated bloodstream infections due to different pathogens) (Price et al., 2002);

- a combination of both (e.g. surgical site infections with *Candida albicans*) (Pertowski et al., 1995).

An alert signal initiates outbreak investigation. This should be focused on epidemiological and microbiological data in order to identify the likelihood of an outbreak. The first evidence that can indicate a potential outbreak is the comparison of baseline and epidemic rates (Barker, Alagoz & Safdar, 2017). At this early stage, it is necessary to create a line list that aggregates relevant information on all potentially affected patients with epidemiological data. Microbiological sampling of pathogens from infection, environment, and patient colonization can then be used for further investigation (Lippmann et al., 2014). This should be conducted in a timely fashion since most microbiological laboratories dispose of their samples after 7 to 14 days. Some infection control departments store pathogens of interest in order to allow for retrospective analysis (Salm et al., 2016). These samples can then be investigated to verify their genetic relatedness using pulsed-field gel electrophoresis or whole genome sequencing (Kampmeier et al., 2017; Sax et al., 2015; Snitkin et al., 2012; Welinder-Olsson et al., 2004). The most cost-effective epidemiological approach is a case–control study, which has the potential to yield information that either reinforces suspected risk factors or leads to undiscovered connections between cases (Moolenaar et al., 2000; Salm et al., 2016).

Outbreak control measures should be directed at the pathogen, the suspected routes of transmission, and its epidemiology within the institution. An orientation on likely transmission pathways and possible control activities can be found in the Centers for Disease Control and Prevention's (CDC) *Guidelines for isolation precautions: preventing transmission of infectious agents in healthcare settings* (Siegel et al., 2007). At the outset of an outbreak, the pathogenesis is often unknown. Given this, control measures are often conducted using a broad, horizontal approach. Therefore, it is imperative for contact to be made with the key clinical players (e.g. physicians, nurses, and department chiefs) as well as to raise awareness in all affected departments (Moolenaar et al., 2000). All players should be informed of the outbreak's course to improve compliance with control measures and ensure that the outbreak serves as a learning opportunity for prevention of future outbreaks under similar conditions.

Antibiotic stewardship

ABS is defined within the goals of an antibiotic stewardship programme (ASP): to optimize clinical outcomes while minimizing unintended consequences of antibiotic use, including toxicity, the selection
of opportunistic pathogens and the emergence of AMR (Dellit et al.,
2007). ABS can be considered as the major tool to achieve responsible
antibiotic use in hospitals. The Driving Reinvestment in Research and
Development and Responsible Antibiotic Use (DRIVE-AB) project
has defined responsible antibiotic use through a RAND-modified
Delphi method, identifying 22 key elements (Monnier et al., 2017).
Together with this definition, a set of 51 inpatient quality indicators
and 12 inpatient quantity metrics was developed using a similar
systematic and stepwise method combining evidence from literature
and stakeholder opinion. A quality indicator reflects the degree to
which antibiotic use is correct or appropriate. In contrast, a quantity metric reflects the volume or the costs of antibiotic use. The
DRIVE-AB process led to multidisciplinary international consensus
on generic quality indicators that can be used globally to assess the
quality of antibiotic use in hospitals. The final set of 12 quantity
metrics includes Defined Daily Dose (DDD) per 1 000 patient days
and Days of Therapy per 1 000 patient days. It is recommended
that antibiotic use should be preferably expressed in at least two
metrics simultaneously (Stanić Benić et al., 2018). The inpatient
quality indicators are very generic and, as with the metrics, should
be refined in order to ensure applicability and measurability across
different health care settings.

Taxonomy of ABS interventions

There have been several systematic reviews on interventions to change
the prescribing behaviour of professionals in hospitals (Davey et al.,
2005; 2013; 2017). Davey et al. performed a critical appraisal using the
Cochrane Effective Practice and Organisation of Care Group (EPOC)
taxonomy (Davey et al., 2017). They appraised all interventions relevant
to improving antibiotic prescribing categorized as persuasive, restrictive,
or structural. The EPOC definitions of the interventions and examples
of the intervention components are given in Table 4.2.

Table 4.2 *EPOC definitions of the interventions and intervention components*

Intervention function	Definition	Intervention components
Education	Increasing knowledge or understanding	Educational meetings; Dissemination of educational materials; Educational outreach
Persuasion	Using communication to induce positive or negative feelings or to stimulate action	Educational outreach by academic detailing or review and recommend change
Restriction	Using rules to reduce the opportunity to engage in the target behaviour (or increase the target behaviour by reducing the opportunity to engage in competing behaviours)	Restrictive
Environmental restructuring	Changing the physical context	Reminders (physical) such as posters, pocket-size or credit card-size summaries or on laboratory test reports; Structural (e.g. new laboratory tests or rapid reporting of results)
Enablement	Increasing means/reducing barriers to increase capability or opportunity	Audit and feedback; Decision support through computerized systems or through circumstantial reminders that were triggered by actions or events related to the targeted behaviour; Educational outreach by review and recommend change

Note: EPOC: Effective Practice and Organisation of Care Group.

Source: Davey et al., 2017.

Evidence on effectiveness of ABS interventions

In their most recent Cochrane review, Davey et al. (2017) concluded that there is high-certainty evidence that interventions are effective in increasing compliance with antibiotic policy and reducing duration of antibiotic treatment. Lower use of antibiotics probably does not increase mortality but does reduce length of stay. Enablement (i.e. increasing means/reducing barriers to increase capability or opportunity) consistently increased the effect of interventions, including those with a restrictive component. Although feedback to health care professionals further increased the effect of an intervention, it was used in only a minority of enabling interventions. Interventions were successful in safely reducing unnecessary antibiotic use in hospitals despite the fact that the majority did not use the most effective behaviour change techniques (Davey et al., 2017).

Schuts et al. (2016a) performed a systematic review and meta-analysis of the evidence on selected ABS objectives, evaluating for the effect of 14 ABS objectives on four predefined patient outcomes: clinical outcome, adverse events, costs, and bacterial resistance rates. The ABS objectives consisted of a mixture of 11 consensus-derived quality indicators of antibiotic use and three additional objectives from the 2007 Infectious Diseases Society of America guidelines on ABS (van den Bosch et al., 2015; Dellit et al., 2007). They identified 145 unique studies with data on nine out of the 14 stewardship objectives. Objective characteristics are summarized in Table 4.3. Overall, the quality of evidence was generally low and heterogeneity between studies was mostly moderate to high. For the objectives empirical therapy according to guidelines, de-escalation of therapy, switch from intravenous to oral treatment, therapeutic drug monitoring, using a list of restricted antibiotics, and bedside consultation, the overall evidence showed significant benefits for one or more of the four outcomes. Guideline-adherent empirical therapy was associated with a 35% relative risk reduction of mortality (RRR) and de-escalation with a RRR of 56% (Schuts et al., 2016a). Evidence of effects was less clear for adjusting therapy according to renal function, discontinuing therapy based on lack of clinical or microbiological evidence of infection, and having a local antibiotic guide. Schuts et al. concluded that for several ABS objectives there is abundant, but low quality, evidence on clinical outcomes, adverse events, costs, and resistance rates in hospitals (Schuts et al., 2016a).

Table 4.3 *Antimicrobial stewardship objectives (145 studies), type of study design and reported outcomes*

Antimicrobial stewardship objective	Number of studies in qualitative synthesis	Type of study design	Outcome data
Empirical therapy according to guidelines	40	All Observational	Mortality Treatment failure LOS Costs
Blood cultures	0	N/A	N/A
Cultures from sites of infection	0	N/A	N/A
De-escalation of therapy[a]	25	1 RCT 24 Observational	Mortality LOS ICU LOS Costs
Adjustment of therapy to renal function	5	All Observational	Mortality ICU LOS Adverse effects Costs
Switch from IV to oral therapy	18	13 RCTs 5 Observational	Mortality Cure or resolution LOS Costs
Documented antibiotic plan	0	N/A	N/A
Therapeutic drug monitoring	16	9 Observational 7 RCTs	Mortality LOS Nephrotoxicity Costs
Discontinuation of antibiotic therapy if infection not confirmed	3	2 RCTs 1 Observational	Mortality ICU LOS Costs
Presence of a local antibiotic guide	1	1 Observational multicentre	Mortality

Table 4.3 *(cont.)*

Antimicrobial stewardship objective	Number of studies in qualitative synthesis	Type of study design	Outcome data
Local guide in agreement with national guidelines	0	N/A	N/A
List of restricted antibiotics	30	29 Observational 1 RCT	Mortality LOS ICU LOS Nosocomial infection rates Costs Resistance rates
Bedside consultation	7	7 Observational	Mortality LOS Costs
Assessment of patients' adherence	0	N/A	N/A

Notes: ªFrom a broad-spectrum to narrower-spectrum antibiotic.

LOS: Length of stay; N/A: Not applicable; RCT: Randomized controlled trial; ICU: intensive care unit; IV: intravenous.

Source: Adapted from Schuts et al., 2016a.

Methodology of ABS interventional studies

This section provides an overview of the methodology used when testing for the effectiveness of antibiotic stewardship programmes. Studies have found overall positive effects but the methodology has not been clearly assessed for external validity or generalization to other populations.

In a narrative review, de Kraker et al. (2017) evaluated for the differences between various study designs in their ability to provide a framework for assessing the quality of evidence for ABS interventions. Relevant literature was identified using a database search of Cochrane and PubMed. The authors found that random time effects and bias can jeopardize the validity of causal inference in ABS research. The most important risks include simultaneously implemented strategies and regression to the mean. Inclusion of homogeneous intervention and control arms, through randomization of the intervention, can limit these risks. However, contamination, that is spill over from the intervention to the control arm, can play an important role for ABS. Therefore, it is recommended that randomization is conducted at the cluster rather than the individual-level. However, it can be challenging to identify enough representative clusters and implementation of a cluster-randomized control trial can be costly. Controlled interrupted time series design has a high validity as well, and is easier to implement, although time-varying confounding should be considered. To detect any unintended consequences, it is crucial to include multiple process, clinical outcome, microbiological and financial measures (de Kraker et al., 2017).

A recent study reviewed published systematic reviews retrieved from Medline to study the evidence base of antibiotic use recommendations and behavioural change interventions (Hulscher & Prins, 2017). It found that most current studies used designs prone to confounding by indication, where participants with less complex or less severe illness may be more likely to have received appropriate antibiotic treatment, which will confound the association between appropriate use and the outcomes tested. Much could be learnt from behavioural sciences. The literature demonstrated that the quality of evidence is low for the positive effects of appropriate antibiotic use in hospital patients. In addition, it found that all types of behavioural change intervention might work. Although effects were positive overall, there were large differences in improvement between studies that tested similar change interventions. Confirming findings elsewhere, the research showed a clear need for studies that use an appropriate study design, i.e. both randomized and controlled, to test for the effectiveness of appropriate antibiotic use in achieving meaningful outcomes (Davey et al., 2017; de Kraker et al., 2017).

ABS guidelines

With growing evidence on the benefit of particular prescription practices, national and international health agencies have issued and regularly updated guidance to address AMR by encouraging appropriate use of antibiotics. In 2014, the CDC recommended that all acute care hospitals in the United States implement ASPs (Fridkin & Srinivasan,

2014). Importantly, the CDC recommends a commitment to stronger leadership to enable dedication of the necessary human, financial, and information technology resources to ASPs. In 2015, the WHO published its *Global action plan on antibiotic resistance* which urges all countries to optimize the use of antibiotic agents (World Health Organization, 2015). In 2016, the Infectious Diseases Society of America issued new antibiotic stewardship guidelines which focused on practical advice for implementing ASPs (Barlam et al., 2016). These replace former, outdated guidelines and focus on specific strategies that are thought to be more beneficial to ensure that the ASP will be effective and sustainable. They recommend that ASPs should tailor interventions based on local issues, resources, and expertise. To ensure this, the guidelines recommend that the ASP is led by physicians and pharmacists and rely on the expertise of infectious diseases specialists. Most recently, ECDC has issued Proposals for EU guidelines on the prudent use of antibiotics in humans in 2017 (European Commission, 2017). Many other EU Member States have issued national guidelines for antibiotic use and ASPs in response to EU-wide calls for action.

Some best ABS practices

National surveillance data on AMR show higher antibiotic use and higher resistance levels in the south and east of Europe compared to the north and west. The Netherlands is an example from the latter, with low consumption and AMR, and a long tradition of antibiotic policies in hospitals. By contrast, national data from Italy show a high antibiotic consumption (ESAC-Net). However, individual hospitals in Italy have started with ASPs focusing on local issues. Box 4.1 shows a selection of exemplary local good antibiotic stewardship practices.

Box 4.1 Two examples of local good antimicrobial stewardship practices

University hospital Modena, Italy

Italy is among the highest consumers of antibiotics and the highest antibiotic resistance rates have been reported (ECDC, 2017). The Clinic of Infectious Diseases, Azienda Ospedaliero-Universitaria, Policlinico di Modena started to expand its antibiotic stewardship

---

Box 4.1 (cont.)

initiative in 2014. The multidisciplinary team reports to the antibiotic and infection prevention committee, which has a mandate from the Board of Directors. Bedini et al. (2016) describe the results of their infectious diseases (ID) consultations in a population of liver cirrhosis patients, with an in-hospital infection rate of more than 30%, mainly caused by Gram-negative microorganisms. Twice a week an ID specialist performed a face-to-face case by case audit with immediate feedback with the gastroenterologist, using (local) guidelines, available diagnostics and the expertise and experience of both physicians. A consensus-based agreement would be reached with the gastroenterologist. Antibiotic consumption and clinical outcome during the first year of the programme were compared with the previous year. The programme resulted in a decrease of antibiotic consumption from 110 to 78 DDD/100 patient days. The greatest impact was observed on carbapenems and quinolones, whose consumption fell by more than 50% without impacting length of stay or in-hospital mortality (Bedini et al., 2016).

National AMR strategy in the Netherlands

The Netherlands has been at the forefront of antibiotic stewardship for more than four decades. The Dutch national Working Party on Antibiotic Policy (SWAB) is funded by the government to conduct antimicrobial surveillance, monitor antibiotic use, and to develop guidelines. In 2006, SWAB developed an online national antimicrobial guide (SWAB-ID) for the prophylaxis and treatment of infectious diseases in hospitals. This concept of an online national antimicrobial guide with local, customizable versions is unique. Use of a local version of this national antimicrobial guide significantly increased both the comprehensiveness and guideline compliance of local antimicrobial policies, and the recommendations were often fed back to the national evidence-based guidelines (Schuts et al., 2016b).

Antibiotic stewardship teams (A-teams), recommended by the Dutch Health Care Inspectorate and the Minister of Health, have been established in every hospital as of 2014. Recent activities include implementing the local antibiotic guide and an observational

---

Box 4.1 (cont.)

pilot study on A-team activities among five Dutch hospitals. The study was conducted to establish a national antibiotic stewardship registry. An assessment was made of the monitoring and documentation of 14 validated stewardship objectives by the A-teams. All A-teams monitored the performance of bedside consultations in *S. aureus* bacteraemia and the prescription of restricted antibiotics. Four fifths of the A-teams could report data on documentation and report on the use of restricted antibiotics. Lack of time and the absence of an electronic medical record system were the main barriers to documentation and reporting (Berrevoets et al., 2017).

Cost–effectiveness of interventions

Governments have limited financial resources. Interventions that are both effective and cost-saving are necessary to reduce the high cost burden of AMR on public health and health system functioning.

A study reviewed the literature on cost–effectiveness of ABS programmes in hospital settings of OECD countries up to June 2014 (Coulter et al., 2015). The type of ABS strategy and the clinical and cost outcomes were evaluated in 36 studies on adult patients. The main ABS strategy implemented was prospective audit with intervention and feedback, followed by the use of rapid diagnostic technology, e.g. rapid polymerase chain reaction (PCR)-based methods or matrix-assisted laser desorption/ionization time-of-flight (MALDI-TOF), for the treatment of bloodstream infections. All but one of the 36 studies reported that ABS resulted in a reduction in pharmacy expenditure. Among 27 studies measuring changes to health outcomes specifically, either no change was reported after the ASP or the additional benefits achieved from these outcomes were not quantified. Only two studies performed a full cost–effectiveness analysis (CEA) (Brown & Paladino, 2010; Scheetz et al., 2009). Both CEAs used a decision-tree model from the hospital perspective and did not evaluate societal costs. Both studies found the interventions to be cost-effective. The earlier study used a model comparing costs and outcomes of bacteraemic patients receiving standard treatment with or without an ASP team consultation (structural intervention). Effectiveness was estimated as quality-adjusted life-years (QALYs) over the lifetime of patients. Incremental cost–effectiveness ratios were calculated to estimate the cost per QALY gained. The later study found that the implementation of rapid testing resulted in improved outcomes for patients. They used data from the literature both from the EU and the USA to inform the model. Rapid PCR testing for MRSA reduced mortality rates while being less costly than empirical therapy in the EU and the USA, even when factoring in a wide range of MRSA prevalence rates and PCR test costs. ABS programmes frequently resulted in a reduction in pharmacy costs. However, there was a lack of consistency in the reported cost outcomes making it difficult to compare the results of the included interventions (Coulter et al., 2015).

The most recent study on the cost–effectiveness and cost–benefits of ASPs summarized the literature from 2000 to 2007 (Naylor et al., 2017). In addition to the CEAs discussed above, it included a CEA that used a Markov model for analysing a bundled ABS strategy (persuasive intervention) conducted in a hospital in Brazil (Okumara et al., 2016). Overall, it concluded that the cost–effectiveness evidence for ABS is severely limited and remains inadequate for investment decision-making. Robust health economics research is needed to enhance the generalizability and usability of cost–effectiveness results.

Conclusions

There are huge challenges in the implementation of infection control and antibiotic stewardship strategies. Increasing levels of antibiotic-resistant pathogens from HAIs indicate the urgent need for early warning systems based on real-time international surveillance. Due to various technological and political barriers, current surveillance systems for HAIs are operating separately on a national level or are laboratory-based with limited clinical and molecular biological data input. Mostly due to technological hurdles, there is a significant delay between data input, analysis, and the report. This diminishes the potential benefits from implementing surveillance programmes, such as monitoring of therapy guidelines, antibiotic formularies, antibiotic stewardship programmes, public health interventions and infection control policies. Following implementation, studies on excess costs of HAIs and ARB-related infections are needed in order to estimate the financial scope of hospital infection control measures and their cost–effectiveness. However, studies on this subject are scarce and often not representative of conditions in diverse settings. As health-care costs are related to the economic circumstances of the particular healthcare system, the comparability between countries is limited.

Any behavioural change intervention in ABS may work in a certain setting for a period of time. However, the evidence on the effectiveness of specific interventions is of rather low quality. The literature shows a clear need for the application of appropriate study designs in a randomized and controlled fashion in order to test the effectiveness of appropriate antibiotic use in achieving meaningful outcomes. The objective would be to identify a set of key interventions with proven effectiveness with results that can be replicated in other settings. Most current studies have used designs prone to confounding by indication. There are many good examples of local practices that could be scaled up to the national level, using insights from the behavioural sciences to select interventions that might work best in a chosen setting. However, a major cause of antibiotic misuse is the insufficient knowledge of prescribing that is rooted in the education system. A suggested approach is to advance the start of education on principles of prudent prescribing towards the undergraduate phase of the medical, pharmacist and nursing curriculum (Pulcini & Gyssens, 2013). It is expected that optimizing the behaviour of professionals requires less effort when attitudes towards prescribing have not yet been shaped or established.

Recommendations

The following recommendations can be made:

- Regarding AMR and HAI surveillance, intensified international collaboration is needed in order to overcome existing barriers to high-quality surveillance.
- Robust data from national surveys are needed to provide useful and comprehensive information to decision-makers in local hospitals.
- To increase the success of educational ABS interventions, education of all professionals should start at the undergraduate level and include medical students, pharmacists and nurses.
- Other stakeholders should be involved to promote responsible antibiotic use in hospitals.

Future directions for research

Although there is a large amount of published literature on interventions to curb AMR in hospitals, some of the relevant outcomes relating to patient health such as patient safety or economics have been neglected. In addition, intervention studies should have more robust designs.

Future research should focus on:

- Targeting treatment and assessing other measures of patient safety, assessing different stewardship interventions, and exploring the barriers and facilitators to implementation. More research is required on unintended consequences of restrictive interventions.
- Robust study designs such as cluster-randomized controlled trials, or interrupted time series including a control arm. A detailed process evaluation should be provided to adequately inform implementation of successful ABS strategies.
- High-quality health economics research on ABS with an appropriate health-economic methodological choice to enhance the generalizability and applicability of cost–effectiveness results.

References

Baker AW, Lewis SS, Alexander BD et al. (2017). Two-phase hospital associated outbreak of Mycobacterium abscessus: investigation and mitigation. Clin Infect Dis. 64(7):902–911.

Barker AK, Alagoz O, Safdar N (2017). Interventions to reduce the incidence of hospital-onset Clostridium difficile infection: An agent-based modeling approach to evaluate clinical effectiveness in adult acute care hospitals. Clin Infect Dis. 66(8):1192–1203.

Barlam TF, Cosgrove SE, Abbo LM et al. (2016). Implementing an antibiotic stewardship program: guidelines by the Infectious Diseases Society of America and the Society for Healthcare Epidemiology of America. Clin Infect Dis. 62(10):e51–77.

Bedini A, De Maria N, Del Buono M et al. (2016). Antimicrobial stewardship in a Gastroenterology Department: Impact on antimicrobial consumption, antimicrobial resistance and clinical outcome. Dig Liver Dis. 48(10):1142–1147.

Berrevoets MA, Ten Oever J, Sprong T et al. (2017). Monitoring, documenting and reporting the quality of antibiotic use in the Netherlands: a pilot study to establish a national antimicrobial stewardship registry. BMC Infect Dis. 17(1):565.

Brown J, Paladino JA (2010). Impact of rapid methicillin-resistant Staphylococcus aureus polymerase chain reaction testing on mortality and cost effectiveness in hospitalized patients with bacteraemia: a decision model. Pharmacoeconom. 28(7):567–575.

Clements A, Halton K, Graves N et al. (2008). Overcrowding and understaffing in modern health-care systems: key determinants in methicillin-resistant Staphylococcus aureus transmission. Lancet Infect Dis. 8(7):427–434.

Conterno LO, Shymanski J, Ramotar K et al. (2007). Impact and cost of infection control measures to reduce nosocomial transmission of extended-spectrum β-lactamase-producing organisms in a nonoutbreak setting. J Hosp Infect. 65(4):354–360.

Coulter S, Merollini K, Roberts JA et al. (2015). The need for costeffectiveness analyses of antimicrobial stewardship programmes: A structured review. Int J Antimicrob Agents. 46(2):140–149.

Damschroder LJ, Banaszak-Holl J, Kowalski CP et al. (2009). The role of the champion in infection prevention: results from a multisite qualitative study. Qual Saf Health Care. 18(6):434–440.

Dancer SJ (2011). Hospital cleaning in the 21st century. Eur J Clin Microbiol Infect Dis. 30(12):1473–1481.

Davey P, Marwick CA, Scott CL et al. (2005). Interventions to improve antibiotic prescribing practices for hospital inpatients. Cochrane Database Syst Rev. 4:CD003543.

Davey P, Brown E, Charani E et al. (2013). Interventions to improve antibiotic prescribing practices for hospital inpatients. Cochrane Database Syst Rev. 4:CD003543.

Davey P, Brown E, Fenelon L et al. (2017). Interventions to improve antibiotic prescribing practices for hospital inpatients. Cochrane Database Syst Rev. 2:CD003543.

de Kraker M (2007). Trends in antimicrobial resistance in Europe: update of EARSS results. Euro Surveill. 12(3):E070315.3.

de Kraker ME, van de Sande-Bruinsma N (2007). Trends in antimicrobial resistance in Europe: update of EARSS results. Euro Surveill. 12(11):3156.

de Kraker ME, Wolkewitz M, Davey PG et al. (2011a). Burden of antibiotic resistance in European hospitals: excess mortality and length of hospital stay associated with bloodstream infections due to Escherichia coli resistant to third-generation cephalosporins. J Antimicrob Chemother. 66(2):398–407.

de Kraker ME, Wolkewitz M, Davey PG et al. (2011b). Clinical impact of antibiotic resistance in European hospitals: excess mortality and length of hospital stay related to methicillin-resistant Staphylococcus aureus bloodstream infections. Antimicrob Agents Chemother. 55(4):1598–1605.

de Kraker ME, Abbas M, Huttner B et al. (2017). Good epidemiological practice: a narrative review of appropriate scientific methods to evaluate the impact of antimicrobial stewardship interventions. Clin Microbiol Infect. 23(11):819–825.

de Kraker ME, Davey PG, Grundmann H et al. (2011c). Mortality and hospital stay associated with resistant Staphylococcus aureus and Escherichia coli bacteremia: Estimating the burden of antibiotic resistance in Europe. PLoS Med. 8(10):e1001104.

Dellit TH, Owens RC, McGowan JE Jr et al. (2007). Infectious Diseases Society of America and the Society for Healthcare Epidemiology of America guidelines for developing an institutional program to enhance antimicrobial stewardship. Clin Infect Dis. 44(2):159–177.

European Centre for Disease Prevention and Control (ECDC) (2010). Antibiotic resistance 2010: global attention on carbapenemase-producing bacteria. Stockholm: European Centre for Disease Prevention and Control. (http://www.eurosurveillance.org/docserver/ fulltext/ eurosurveillance/15/46/art19719-en.pdf?expires=1511794526 &id=id&accname=guest&checksum=797316D1C4E88391081819 C8A61E9267, accessed 06 September 2018).

European Centre for Disease Prevention and Control (ECDC) (2013). Point prevalence survey of healthcare-associated infections and antibiotic use in European acute care hospitals, 2011–2012. Stockholm: European Centre for Disease Prevention and Control. (https://ecdc.europa.eu/sites/portal/ files/media/en/publications/Publications/ healthcare-associated-infections-antimicrobial-use-PPS.pdf, accessed 06 September 2018).

European Centre for Disease Prevention and Control (ECDC) (2016). Technical Document: Point prevalence survey of healthcare-associated infections and antimicrobial use in European acute care hospitals Protocol version 5.3 2016–2017. Stockholm: European Centre for Disease Prevention and Control. (https://ecdc.europa.eu/sites/portal/files/media/en/publications/ Publications/PPS-HAI-antimicrobial-use-EU-acute-care-hospitals-V5-3.pdf, accessed 06 September 2018).

European Centre for Disease Prevention and Control (ECDC) (2017). Antimicrobial resistance surveillance in Europe 2016. Stockholm: European Centre for Disease Prevention and Control. (https://ecdc.europa.eu/en/ publications-data/antimicrobial-resistance-surveillanceeurope-2016, accessed 06 September 2018).

European Commission (2017). EU Guidelines for prudent use of antimicrobials in human health. Brussels: European Commission. (https://ec.europa.eu/

health/amr/sites/amr/files/amr_guidelines_prudent_use_en.pdf, accessed 06 September 2018).

Fridkin SK, Srinivasan A (2014). Implementing a strategy for monitoring inpatient antimicrobial use among hospitals in the United States. Clin Infect Dis. 58(3):401–406.

Gagliotti C, Balode A, Baquero F et al. (2011). Escherichia coli and Staphylococcus aureus: bad news and good news from the European Antibiotic Resistance Surveillance Network (EARS-Net, formerly EARSS), 2002 to 2009. Euro Surveill. 16(11):19819.

Gastmeier P, Geffers C, Brandt C et al. (2006). Effectiveness of a nationwide nosocomial infection surveillance system for reducing nosocomial infections. J Hosp Infect. 64(1):16–22.

Gastmeier P, Sohr D, Schwab F et al. (2008). Ten years of KISS: the most important requirements for success. J Hosp Infect. 70:11–16.

Gastmeier P, Schröder C, Behnke M et al. (2014). Dramatic increase in vancomycin-resistant enterococci in Germany. J Antimicrob Chemother. 69(6):1660–1664.

Graves N, Plowman R, Roberts JA (2001). The epic project: developing national evidence-based guidelines for preventing healthcare associated infections. J Hosp Infect. 48(4):320–321.

Halton K, Graves N (2007). Economic evaluation and catheter-related bloodstream infections. Emerg Infect Dis. 13(6):815–823.

Harris AD, Nemoy L, Johnson JA et al. (2004). Co-carriage rates of vancomycin-resistant Enterococcus and extended-spectrum betalactamase-producing bacteria among a cohort of intensive care unit patients: implications for an active surveillance program. Infect Control Hosp Epidemiol. 25(2):105–108.

Haustein T, Gastmeier P, Holmes A et al. (2011). Use of benchmarking and public reporting for infection control in four high-income countries. Lancet Infect Dis. 11(6):471–481.

Huang SS, Septimus E, Kleinman K et al. (2013). Targeted versus universal decolonization to prevent ICU infection. N Engl J Med. 368(24):2255–2265.

Hübner C, Hübner NO, Hopert K et al. (2014). Analysis of MRSA-attributed costs of hospitalized patients in Germany. Eur J Clin Microbiol Infect Dis. 33(10):1817–1822.

Hulscher ME, Prins JM (2017). Antibiotic stewardship: does it work in hospital practice? A review of the evidence base. Clin Microbiol Infect. 23(11):799–805.

Kaier K, Mutters N, Frank U (2012). Bed occupancy rates and hospital-acquired infections – should beds be kept empty? Clin Microbiol Infect. 18(10):941–945.

Kampmeier S, Knaack D, Kossow A et al. (2017). Weekly screening supports terminating nosocomial transmissions of vancomycin-resistant enterococci on an oncologic ward – a retrospective analysis. Antimicrob Resist Infect Contr. 16(1):48.

Leistner R, Piening B, Gastmeier P et al. (2013). Nosocomial infections in very low birthweight infants in Germany: current data from the National Surveillance System NEO-KISS. Klinische Padiatrie. 225(2):75–80.

Lippmann N, Lübbert C, Kaiser T et al. (2014). Clinical epidemiology of Klebsiella pneumoniae carbapenemases. Lancet Infect Dis. 14(4):271–272.

Longtin Y, Sax H, Allegranzi B et al. (2011). Videos in clinical medicine: Hand hygiene. N Engl J Med. 364(13):e24.

Marshall C, Spelman D (2007). Is throat screening necessary to detect methicillin-resistant Staphylococcus aureus colonization in patients upon admission to an intensive care unit? J Clin Microbiol. 45(11):3855.

Monnier AA, Eisenstein BI, Hulscher ME et al. (2018). Towards a global definition of responsible antibiotic use: results of an international multidisciplinary consensus procedure. J Antimicrob Chemother. 73(6):3–16.

Moolenaar RL, Crutcher JM, San Joaquin VH et al. (2000). A prolonged outbreak of Pseudomonas aeruginosa in a neonatal intensive care unit: did staff fingernails play a role in disease transmission? Infect Control Hosp Epidemiol. 21(2):80–85.

Naylor NR, Zhu N, Hulscher M et al. (2017). Is antibiotic stewardship cost-effective? A narrative review of the evidence. Clin Microbiol Infect. 23(11):806–811.

OECD (2014). Paying providers for healthcare. Paris: Organisation for Economic Co-operation and Development. (http://www.oecd.org/els/health-systems/paying-providers.htm, accessed 06 September 2018).

Okumara LM, Riveros BS, Gomes-da-Silva MM et al. (2016). A cost–effectiveness analysis of two different antibiotic stewardship programs. Brazil J Infect Dis. 20(3):255–261.

Pertowski CA, Baron RC, Lasker BA et al. (1995). Nosocomial outbreak of Candida albicans sternal wound infections following cardiac surgery traced to a scrub nurse. J Infect Dis. 172(3):817–822.

Pittet D, Hugonnet S, Harbarth S et al. (2000). Effectiveness of a hospitalwide programme to improve compliance with hand hygiene. Lancet. 356(9238):1307–1312.

Pittet D, Allegranzi B, Sax H et al. (2006). Evidence-based model for hand transmission during patient care and the role of improved practices. Lancet Infect Dis. 6(10):641–652.

Price CS, Hacek D, Noskin GA et al. (2002). An outbreak of bloodstream infections in an outpatient hemodialysis center. Infect Control Hosp Epidemiol. 23(12):725–729.

Pulcini C, Gyssens IC (2013). How to educate prescribers in antimicrobial stewardship practices. Virulence. 4(2):192–202.

Salm F, Deja M, Gastmeier P et al. (2016). Prolonged outbreak of clonal MDR Pseudomonas aeruginosa on an intensive care unit: contaminated sinks and contamination of ultra-filtrate bags as possible route of transmission? Antibiot Resist Infect Cont.5:53.

Sax H, Allegranzi B, Uçkay I et al. (2007). "My five moments for hand hygiene": a user-centred design approach to understand, train, monitor and report hand hygiene. J Hosp Infect. 67(1):9–21.

Sax H, Bloemberg G, Hasse B et al. (2015). Prolonged outbreak of Mycobacterium chimaera infection after open-chest heart surgery. Clin Infect Dis. 61(1):67–75.

Scheetz MH, Bolon MK, Postelnick M et al. (2009). Cost-effectiveness analysis of an antibiotic stewardship team on bloodstream infections: a probabilistic analysis. J Antimicrob Chemother. 63(4):816–825.

Schröder C, Schwab F, Behnke M et al. (2015). Epidemiology of healthcare associated infections in Germany: Nearly 20 years of surveillance. Int J Med Microbiol. 305(7):799–806.

Schuts EC, Hulscher MEJL, Mouton JW et al. (2016a). Current evidence on hospital antimicrobial stewardship objectives: a systematic review and meta-analysis. Lancet Infect Dis. 16(7):847–856.

Schuts EC, van den Bosch CM, Gyssens IC et al. (2016b). Adoption of a national antimicrobial guide (SWAB-ID) in the Netherlands. Eur J Clin Pharmacol. 72(2):249–252.

Shepard J, Ward W, Milstone A et al. (2013). Financial impact of surgical site infections on hospitals: the hospital management perspective. JAMA Surgery. 148(10):907–914.

Siegel JD, Rhinehart E, Jackson M et al. (2007). Guideline for isolation precautions: preventing transmission of infectious agents in healthcare settings. Atlanta: US Centers for Disease Control and Prevention. (https://www.cdc.gov/infectioncontrol/pdf/guidelines/isolation-guidelines.pdf, accessed 06 September 2018).

Snitkin ES, Zelazny AM, Thomas PJ et al. (2012). Tracking a hospital outbreak of carbapenem-resistant Klebsiella pneumoniae with whole-genome sequencing. Sci Transl Med. 4(148):148ra116.

Stanić Benić M, Milanic R, Monnier AA et al. (2018). Metrics for quantifying antibiotic use in the hospital setting: results from a systematic review and international multidisciplinary consensus procedure. J Antimicrob Chemother. 73(6):50–58.

Tacconelli E, Cataldo MA, Dancer SJ et al. (2014). ESCMID guidelines for the management of the infection control measures to reduce transmission of multidrug-resistant Gram-negative bacteria in hospitalized patients. Clin Microbiol Infect. 20 Suppl(1):1–55.

Tacconelli E, Sifakis F, Harbarth S et al. (2017). Surveillance for control of antibiotic resistance. Lancet Infect Dis. 18(3):99–106.

Tankovic J, Legrand P, De Gatines G et al. (1994). Characterization of a hospital outbreak of imipenem-resistant Acinetobacter baumannii by phenotypic and genotypic typing methods. J Clin Microbiol. 32(11):2677–2681.

Tschudin-Sutter S, Lucet JC, Mutters NT et al. (2017). Contact precautions for preventing nosocomial transmission of extended-spectrum beta lactamase-producing Escherichia coli: A point/counterpoint review. Clin Infect Dis. 65(2):342–347.

van den Bosch CM, Geerlings SE, Natsch S et al. (2015). Quality indicators to measure appropriate antibiotic use in hospitalized adults. Clin Infect Dis. 60(2):281–291.

Weintrob AC, Roediger MP, Barber M et al. (2010). Natural history of colonization with gram-negative multidrug-resistant organisms among hospitalized patients. Infect Control Hosp Epidemiol. 31(4):330–337.

Welinder-Olsson C, Stenqvist K, Badenfors M et al. (2004). EHEC outbreak among staff at a children's hospital – use of PCR for verocytotoxin detection and PFGE for epidemiological investigation. Epidemiol Infect. 132(1):43–49.

World Health Organization (2015). Global action plan on antibiotic resistance. Geneva: World Health Organization. (http://apps.who.int/iris/bitstream/10665/193736/1/9789241509763_eng.pdf, accessed 06 September 2018).

Zimlichman E, Henderson D, Tamir O et al. (2013). Health care-associated infections: a meta-analysis of costs and financial impact on the US health care system. JAMA Intern Med. 173(22):2039–2046.

Zingg W, Holmes A, Dettenkofer M et al. (2015). Hospital organisation, management, and structure for prevention of health-care-associated infection: a systematic review and expert consensus. Lancet Infect Dis. 15(2):212–224.

5 | Tackling antimicrobial resistance in the food and livestock sector

JEROEN DEWULF, SUSANNA STERNBERG-LEWERIN, MICHAEL RYAN

Introduction

As the world population continues to grow, the demand for livestock and livestock products also rises, resulting in further increases in large-scale intensive livestock production to meet this increased demand. Accompanying this intensification in many countries is a rise in the use of antibiotics in the production system. This is particularly the case for emerging economies, particularly large animal-producing countries. For most Organisation for Economic Co-operation and Development (OECD) countries, however, the use of antimicrobials in livestock production is falling, as the traditional livestock production systems evolve and alternative approaches to disease management are adopted.

The growing resistance of microbes to the commonly used antimicrobials is a serious concern for human and animal health and, thus, for policy-makers in many countries. Moreover, it also raises important questions in relation to food safety, food security, trade and market access for livestock and livestock products. Globally, antibiotics are widely used in livestock production for a range of purposes, with the bulk used in the high-density intensive livestock production systems. The global antimicrobial use (AMU) in livestock can be divided into therapeutic, metaphylactic, prophylactic, and growth promotion. Antibiotic use is strictly under veterinary prescription in most OECD countries, but in many parts of the world veterinary drugs are available over the counter in pharmaceutical stores. In low-income countries, weaknesses in the legislative and veterinary infrastructure often present a challenge to the regulation of the access to veterinary drugs. The World Organisation for Animal Health (OIE) has worked to strengthen veterinary services and more recently to encourage global reporting of antimicrobial use in animals (http://www.oie.int/scientific-expertise/veterinary-products/antimicrobials).

Therapeutic use refers to the use of antibiotics to treat clinically diseased animals; whereas metaphylactic use involves treatment of entire groups of animals when some individuals in the group are diseased to avoid further spread of the infection. Prophylactic use is generally defined as preventive antibiotic use to avoid clinical problems (e.g. at a certain stage in the production cycle). Antimicrobial growth promotion means regular inclusion of subtherapeutic doses of antibiotics in feed with the aim of improving the feed conversion and growth rate of the animals. Antibiotics are often used as a regular and systematic input in intensive livestock production, as the productivity and financial benefits are perceived to outweigh the costs. They therefore have important implications for output, which, in turn, affects commodity markets and trade in livestock and livestock products.

The use of antibiotics in animal production has important implications not only for animal health and welfare, but also for food safety and food security at the global level. While comprehensive information and data on the productivity impact of antibiotics are sparse, data collection is improving as more resources are being employed in many countries to monitor AMU, as well as the growth and health impacts of this use. While antimicrobials are an important input in disease management in some modern livestock production systems, their use inevitably results in selection for antimicrobial resistance (AMR), which raises serious concerns that need to be addressed. In addition to the excessive AMU, other drivers, such as antimicrobial waste from farms or manufacturers, may also contribute to the rise in AMR. Besides AMU, other factors, such as the use of heavy metals (such as zinc oxide given to prevent weaning diarrhoea in piglets), may result in co-selection of resistance traits. In practice, an increasing number of studies have shown that AMU in humans (Charbonneau et al., 2006; Costelloe et al., 2010; Sun, Klein & Laxminarayan, 2012) is the main driver for AMR in human bacteria, whereas AMU in animals (Burow et al., 2013; Hammerum et al., 2014; Simoneit, Burow & Tenhagen, 2015; Chantziaras et al., 2014) is the main driver for the development of AMR in animal bacteria. Yet there is also evidence for spill over of resistance from animals to humans and vice versa (Cabello, 2006; Crombé et al., 2013; Liu et al., 2016; Madec et al., 2017).

Limiting the use of antimicrobials in livestock production is a challenge due to different regulatory systems, definitional issues, measurement methods, surveillance and monitoring challenges. Moreover, there is growing debate on the perceived short-term private benefits of AMU, primarily to livestock producers, versus the longer-term social costs of AMR on human health, the environment, animal health, and food production. At the global level, estimates have been made of the potential economic costs associated with the rise in AMR in the medium to long run (Laxminarayan, Van Boeckel & Teillant, 2015; Adeyi et al., 2017). Most of these estimates only relate to the impact on human health, with little empirical evidence of the impact on livestock production and, consequently on food supplies. Nonetheless, there are increasing numbers of reports on therapy failures in animal diseases linked to growing resistance levels.

This chapter reviews the current situation in terms of antibiotic use in modern livestock production. The core issues related to the impact of antibiotics in treating disease outbreaks and the productivity affects are explored. Moreover, the problems faced in measuring AMU and AMR continue to be rather contentious due to the lack of a harmonized global approach. The complexity of the transmission pathways between animals, the environment and humans remains a challenge to the better understanding of the means and speed of transmission of resistant pathogens and how long these pathogens remain viable in the environment. The final section identifies some pragmatic interventions that have been successfully adopted at the farm level to reduce AMU.

Current state of knowledge on antimicrobial use and antimicrobial resistance in livestock production and the food-chain

AMU and AMR in livestock production

Addressing the issue of the use of antimicrobials in meat production is complex because they are used to achieve both a health and a productivity benefit in livestock-producing farms. The multipurpose objectives of AMU in agriculture include therapeutic, metaphylactic, prophylactic use, and use for growth promotion. Of these, antimicrobial growth promoters (AGP) are clearly nontherapeutic while prophylactic and metaphylactic use falls somewhere in between nontherapeutic and therapeutic use. Some animal categories in intensive livestock production are particularly susceptible to infections, and although routine antibiotic treatment of such animal groups should be classified as prophylactic (or metaphylactic) use, it is often regarded as therapeutic, demonstrating the challenges when using these definitions in policy-making.

Box 5.1 Challenges in categorizing antibiotics as therapeutic, metaphylactic or prophylactic

In some management systems, post-weaning diarrhoea in piglets has been regarded as inevitable without routine treatment with antibiotics, but improved management has proven that this disease can be prevented. Some animal production systems have a turnover rate that presents a challenge for prophylactic tools such as vaccines because of the time needed to develop an effective immune response. Metaphylactic use of ionophore antibiotics to prevent coccidiosis has been routinely applied in broiler production in large parts of the world. New vaccines and management optimization has allowed for a substantial reduction of this use in many countries. However, although ionophores are mainly used to prevent a parasitic infection (coccidiosis) they also prevent necrotic enteritis in poultry (a multifactorial disease induced by the presence of the bacterium *Clostridium perfringens*) and both of these diseases must be managed for a successful reduction of ionophore usage.

While the principal role of AMU in food animals should be therapeutic, in reality use has been substantially driven by the objective of improving farm productivity and income. Evaluating the impacts of antibiotics on animal productivity is difficult due to the relatively limited number of studies on the different food animal species. High antibiotic use is often related to poor management or health failures on the farm. The key question is whether the use of antibiotics could be replaced by better husbandry, management standards and production systems, and at what cost? Historically, the use of antibiotics in livestock production has been closely correlated to the size of the livestock population and the intensity of the production system in a country. Highly intensive animal production systems have tended to use more antibiotics than less intensive systems. However, during recent decades the sector has seen the rapid development of intensive production systems with higher biosecurity measures, improved husbandry and management, which together have led to a reduction in AMU in many countries (Postma et al., 2016; Laanen et al., 2014).

A 2012 study noted that over two fifths of all feedlot cattle and over four fifths of hogs in the USA were given antibiotics in their feed rations

(Landers et al., 2012). Another study estimated that food and agriculture production accounts for the bulk of antimicrobials consumed in the USA, estimated at over 70% of total consumption (Laxminarayan, Van Boeckel & Teillant, 2015). The study also estimated the global volume of antibiotics consumed in agriculture at 63 000 tonnes in 2010, and noted that this would rise to 106 000 tonnes by 2030 if no changes are made in the use of antibiotics. The authors attributed two thirds of this increase to a rise in the number of food animals and one third to more intensive livestock production systems. It also noted that four countries, namely China, the United States, India and Brazil, account for almost 50% of total global consumption, and that this would remain unchanged in the coming decade.

The study concluded that the consumption of antibiotics is closely related to the livestock population, with the highest consumption in countries which have a high concentration of industrial pig, poultry and cattle enterprises. These projections assume that no changes are introduced in the way antibiotics are used in animal production in the near future. However, several European countries have recently shown that very substantial decreases in use can be achieved in a short period of time without negative effects on animal health and production as long as they are accompanied by improved management and biosecurity practices. Also, the emergence of private initiatives and labels such as "No antibiotics ever" may have led to a decrease in antibiotic use in the US broiler chicken industry and a fall in the total consumption of antimicrobials. In several large animal-producing Asian countries important policy changes have recently been made regarding the use of antimicrobial growth promoters. Therefore, it is reasonable to expect that the predictions referred to above may be unduly pessimistic.

Aquaculture is one of the fastest growing food production sectors, and is regarded as an important part of the solution to global food insecurity. However, similar problems with infectious diseases and overuse of antibiotics, as seen in intensive terrestrial animal production systems, have been experienced in aquaculture. Prophylactic use of antimicrobials is common in many regions (Cabello, 2006; Done, Venkatesan & Halden, 2015). However, in aquaculture there has been some success in reducing antimicrobial use by the use of preventive measures such as vaccination; for example, the aeromonas vaccine in salmon production (Gulla et al., 2016). Figure 5.1 summarizes the pathways of transmission of resistant bacteria in the environment.

Figure 5.1 Summary of the pathways of transmission of resistant bacteria between animals, humans and the environment

Note: The above image depicts the pathways of transmission of resistant bacteria between animals, humans and the environment. Such as; dissemination through water sanitation systems (1), the application of manure to fields with cultivated crops (2), which then leads to antibiotic-resistant bacteria developing on plants (3). The uptake of resistant bacteria through the food-chain (4) or within the meat products harbouring resistant bacteria (5). Water distribution systems can also spread resistant bacteria (6). Wildlife, insects and other bugs are also carriers of resistant bacteria (7). Lastly, tourism, migrations and trade (8) are drivers of spreading resistant bacteria across borders.

Source: bioMérieux, 2016.

There is growing alarm at the rise in AMR and the potential consequences for food production, animal and human health. The use, overuse and misuse of antibiotics drives an increase in AMR and gives rise to serious technical difficulties when treating animal diseases. In food animal production, the rise in AMR not only increases the risk of animal mortality, but the inability to treat resistant infections also reduces animal performance, thus reducing the economic returns in agriculture and the food system, and potentially higher food prices for consumers. Current research indicates that livestock production accounts for more than two thirds of global antibiotics consumption. However, with the implementation of the WHO's Global Action Plan on Antimicrobial Resistance (World Health Organization, 2015), and the greater global awareness of the risks associated with increasing AMR, the use of antibiotics is likely to decline in the coming years. A more prudent approach to antibiotics consumption is necessary to slow the pace at which resistance develops.

Productivity gains appear to be declining

Prevention and management of animal diseases are critical in modern livestock production. An outbreak of an infectious disease on the farm not only reduces productivity and output, but also increases the costs of treating the animals. In modern and sustainable livestock production the focus should be on disease prevention. Only if this fails should antimicrobials be used. The productivity impact arising from AMU varies substantially by species and stage of growth of each species. And, as the productivity gain from AMU varies with health status and management, cost–benefit estimates are needed that take into account long-term costs of AMR to producers.

Box 5.2 Examples of responses to antibiotic reduction

The Netherlands and Belgium have recently reduced their antibiotic use in animal production substantially and demonstrated that this can have an almost immediate effect on lowering the levels of resistance in animal production (Dorado-García et al., 2016; Callens et al., 2017). However, for some resistance traits, the process is more complex. If resistance involves a "cost" for the bacteria, i.e. slows their growth, or is located on a genetic element that is easily lost, removal of the selective pressure exerted by AMU will lead to loss of the resistance. If the resistance trait does not impair bacterial growth, or is linked to other genes that are needed in the current microenvironment, resistant bacteria will not be at a disadvantage when the selective pressure is removed and resistance may remain at high levels. This illustrates that although resistance selection is not an irreversible process it must be slowed down immediately and urgent intervention is called for.

Earlier studies on the production impacts of antibiotics given in the feed as growth promoters indicate productivity gains ranging from 1% to double digits, depending on factors such as nutrition, breeding, housing, sanitation, as well as husbandry and management practices. However, recent studies have concluded that the productivity benefits from the routine use of low levels of antibiotics in the feed have declined due in part to the adoption of modern production and management practices (Laxminarayan, Van Boeckel & Teillant, 2015). Hence, poor management systems have benefited from the use of AGPs but they should have no place in modern animal production as AMR is promoted by their use.

However, several factors can influence the performance including the species (pigs, poultry, cattle), age of the animal, nutrition, breed, as well as the production system and management practices. There is evidence to suggest that AGPs have no effect when fed to germ-free animals (Swedish Ministry of Agriculture, 1997). The intestinal characteristics of germ-free animals resemble the effects reported from AGP use. It has been proposed that most of the effect of AGPs is due to suppression of intestinal microbes that induce host immune responses that are detrimental to efficient growth (Broom, 2017). In management systems with good hygiene and improved animal health, production performance is already optimized and the net benefit of AGPs is very doubtful.

Sweden was the first OECD country (in 1986) to ban the use of antibiotics as a growth promoter in animal feed. This was followed by several other countries and the EU banned the use of antibiotic growth promoters in animal feed for all EU Member States in 2006 (Regulation 1831/2003/EC). In contrast to some expectations, this ban has not resulted in a substantial decline in food animal production in Europe. While the first ban in Sweden led to some initial animal health problems that had to be addressed by improved management and disease prevention (Swedish Ministry of Agriculture, 1997), lessons learnt from the Swedish experience helped other EU/EEA countries to cope with the subsequent union-wide ban that was applied more gradually. In Denmark and the Netherlands a shift towards therapeutic AMU was also observed after the ban of antimicrobial growth promoters. However, this increase turned out to be only temporary and was even non-existent in Norway (Bos et al., 2013; Grave et al., 2004; 2006).

In the USA, the use of medically important antibiotics for animal growth promotion has been banned only since 2017. Several other OECD countries which have banned the use of antimicrobials for growth

---

Box 5.3 Reduction of antibiotic consumption through direct guidelines

The Netherlands has implemented clear reduction targets and a range of measures such as a ban on in-feed mixing of antimicrobials, herd level monitoring of use, increased awareness building, and strict regulations on the farmer–veterinarian relationship. This has resulted in a 56% drop in consumption of antibiotics in agriculture between 2007 and 2012, without any serious adverse effects on animal welfare or on the profitability of the farms (Speksnijder et al., 2015). Countries such as Belgium, France, Germany and the UK have implemented initiatives, including the setting of reduction targets, which also show promising reductions in antibiotic use.

---

promotion in the last decade include Mexico, New Zealand and the Republic of Korea, while other countries such as Indonesia, Viet Nam, and China have recommended the gradual removal of antibiotics as a growth stimulant.

Given the risk related to AMU and resistance selection, several European countries, including Denmark, Sweden, Belgium and the Netherlands, have introduced strict limits on the consumption of antibiotics on livestock farms. This has contributed to a significant fall in antibiotics usage in these countries without a substantial negative impact on production.

These initiatives show that substantial reductions in AMU are possible and that these initiatives should be focused on alternative disease prevention actions (e.g. improving biosecurity and animal husbandry). Limiting the increase in AMR requires a focus on both the demand for antibiotics and the supply of antibiotics for animal use. It is important that, first, the need for antimicrobials is reduced by focusing on disease prevention and improved production. In a second stage, access to necessary antibiotics to treat infectious diseases should be maintained, while at the same time eliminating the overuse and misuse of antibiotics in animal production.

Measuring AMU and AMR in animal production

There are enormous global challenges in measuring AMU in animal production, due to lack of resources, expertise and understanding of the adverse consequences of increasing levels of AMR. These difficulties have limited the availability of reliable and comparable data across species and across countries. As a consequence, international organizations have encouraged the collection of AMU data to manage and minimize the further development of AMR (World Health Organization, 2015; World Organisation for Animal Health, 2016). The newly proposed EU Regulation on veterinary medicinal products regulates the collection of data on AMU in Member States and requires that such data should be comparable, compiled on an EU level, and published annually. The OIE, supported by the FAO and WHO within the Tripartite Collaboration on AMR, has taken the lead in building a global database for antimicrobial agents intended for use in animals (OIE, 2018).

While good progress has been made in recent years, the lack of comprehensive data has limited the development of alternative interventions to antibiotics in animal production. Comparable data are needed for benchmarking and assessing interventions to reduce AMU. EU legislation only allows AMU based on veterinary prescription, but sales of antibiotics are not regulated in most parts of the world. Within the EU, some Member States allow veterinary sales of antibiotics while others forbid veterinarians to make a profit from supplying medicines. Data on AMU may be held by various actors, such as the pharmaceutical industry, pharmacies and veterinary clinics. However, the difficulties lie not only in collecting data from multiple sources, but in assessing the actual use in regards to the number of treated animals of different species. Even when prescription data are available, these do not always provide enough detail to account for the exact number of daily doses, mainly due to large differences in body weight between animals of different age categories and different dosages for different routes (and formulae) of administration (Collineau et al., 2016). The European Surveillance of Veterinary Antimicrobial Consumption project aims to develop a system for collection of data per animal species and to establish technical units of measurement (EMA, n.d.).

Although efforts are made to collect and assess data on AMU in many parts of the world, challenges remain elsewhere. In systems where farmers buy feed without knowing the exact contents, and where antibiotic substances are only listed as "additives" on feed labels, much AMU may go unnoticed. In most parts of South-East Asia, Africa and South America, intensive production systems in general, and aquaculture in particular, there is a lack of specific information on AMU, but it is suspected to be quite substantial (Krishnasamy et al., 2015).

When it comes to AMR, the challenges are different. Most countries with ongoing or planned monitoring of AMU already have surveillance for AMR in place, but lack of laboratory capacity is one of the major problems globally. In regions where resources are lacking, available data are scarce, sporadic and usually non-validated. Even in regions where animal producers can afford diagnostics and these are available, data on AMR may be sporadic and difficult to compare when reliant on clinical samples alone. While such samples are valuable, they cannot replace systematic monitoring of AMR in indicator bacteria. This is needed for comparison between production categories, animal species, regions and over time, as a basis for interventions and benchmarking.

Box 5.4 Surveillance of antibiotic consumption and sales data

The US Department of Agriculture (USDA), in close collaboration with the US Food and Drug Administration, has also initiated projects that aim to assess AMU at the farm level, but data collected at the national level are not yet available. In Canada, information on AMU in animals is provided by the Canadian Animal Health Institute on a voluntary basis, based on sales data from companies, while mandatory reporting of sales data and collection of more detailed data on antimicrobial consumption have been proposed in Australia. The European Surveillance of Veterinary Antimicrobial Consumption collects information on how antimicrobials are used in animals across the European Union (EMA, n.d.). The ECDC, in conjunction with the European Medicines Agency (EMA) and the European Food Safety Authority (EFSA), also undertakes joint analysis of the consumption of antimicrobials and the occurrence of antimicrobial resistance in bacteria from humans and food-producing animals (ECDC/EFSA/EMA, 2017). Several countries have initiated sector-driven initiatives on measuring AMU. The recently established AACTING[1] network has compiled all currently available systems for measuring AMU in animals at herd level and has identified at least 24 farm-level data collection systems from 15 countries in Europe and Canada.

1 Network on quantification of veterinary antimicrobial usage at herd level and analysis, communication and benchmarking to improve responsible usage (http:// www.aacting.org).

As an example, in Europe, this type of AMR monitoring of indicator bacteria is undertaken by the European Food Safety Authority (EFSA) and the ECDC, resulting in an annual report on the presence of AMR in zoonotic and commensal bacteria originating from food-producing animals and animal products (EFSA/ECDC, 2018). Passive surveillance of AMR, based on clinical samples alone, will provide some insight into the current clinical problems and certain animal health threats due to AMR, but is less useful than active surveillance for monitoring trends and making comparisons between settings.

Risk assessment aspects of AMR

Risk assessment forms the basis for planned interventions in animal production (risk management). This is not always straightforward. While the association between AMU and AMR is indisputable, it is often difficult to quantify as there are so many factors, such as dose and duration of treatment, route of administration, co- and cross-resistance selection effects, all influencing the direct association between the use of one specific antimicrobial and the rise of one specific type of resistance. However, there is a general agreement that AMU must be reduced in livestock production and, hence, risk assessment could focus on AMU. In order to appreciate the risk of AMR (and AMU), producers must have knowledge and awareness. The levels of these vary and, therefore, the risk profile of producers will differ according to the production system, species, age of the producer, geography, cultural norms and behaviour, as well as the potential economic aspects.

The structure of the production system plays an important role in determining the behaviour of producers with respect to the threats posed by AMR. Producers in the early stages of the animal life-cycle may regard the risks of not using antibiotics as more important than consumer perceptions while producers in the later stages are more dependent on consumer confidence and more directly affected by withdrawal periods and subsequent losses due to treatments. The relationship between knowledge, attitude and behaviour is complex, as illustrated by studies on farmers' implementation of biosecurity measures (Racicot et al., 2012; Laanen et al., 2014; Kristensen et al., 2011). As reduction of on-farm AMU is dependent on disease prevention measures, these results are relevant for the issue of AMU reduction.

Most antibiotics are used both in humans and in animals

Research has shown that over 20 of the 27 different classes of antibiotics are used in both animals and humans. There is growing concern in relation to livestock production over the use of last-resort antibiotics for humans, such as colistin, as these are increasingly required for use in humans as AMR spreads. Many drugs that had been discarded for human use due to toxicity issues have been used in animals as growth promoters or as prophylaxis or therapy for enteric infections, and now are coming back as last-resort drugs in human medicine. The earliest example is avoparcin, which was banned as an AGP when selection for vancomycin resistance due to cross-resistance was reported (Swedish Ministry of Agriculture, 1997). Other AGPs that were previously regarded as irrelevant for human medicine but that have been discussed as potential candidates for last-resort drugs are avilamycin and flavomycin.

Resistance to colistin, a polymyxin substance widely used in pig and poultry production, was previously reported exclusively due to chromosomal mutations. In 2015, Chinese investigations into increased prevalence of colistin-resistant *Escherichia coli* from pigs revealed a resistance gene located on a plasmid (Liu et al., 2016). Hence, following the discovery of this transferable colistin resistance, the European Medicines Agency published targets for reduction of colistin use in animals in EU Member States, as well as a reclassification of colistin as a medicine "reserved for treating infections in animals for which no effective alternative treatments exist". Countries such as China and Brazil have taken targeted measures to reduce the use of colistin. This example demonstrates the gradual transition from regarding AMU in humans and animals as separate, to a realization that this is indeed a One Health issue.

Transmission of AMR between livestock, the environment and humans

In the past there were populations that were extremely isolated, but globalization means that all parts of the world are now interconnected. Humans and animals (domestic animals as well as wildlife) continuously interact, both with each other and with the specific environment or ecosystem they inhabit. Moreover, in many parts of the world antimicrobials are used (in humans and animals), all potentially selecting for AMR.

Therefore, it is clear that the growth of AMR cannot be addressed by simply acting on one element. What happens in human medicine has an impact on the environment and the bacterial flora in animals. Similarly, what happens in veterinary medicine influences the bacterial flora in humans. This becomes even more obvious in ecosystems where intensive farming (of animals and crops) is combined with a dense population, providing the ideal circumstances for a dynamic exchange of bacteria and resistance genes. Figure 5.2 shows the potential routes for the exchange of resistant bacteria between animals and humans, and vice versa. It is important to emphasize that the exchange of resistant traits may go in both directions, from animals to humans as well as from humans to animals.

In the exchange between animals and humans, three types of transfer mechanisms can be distinguished. First, resistant traits (bacteria or

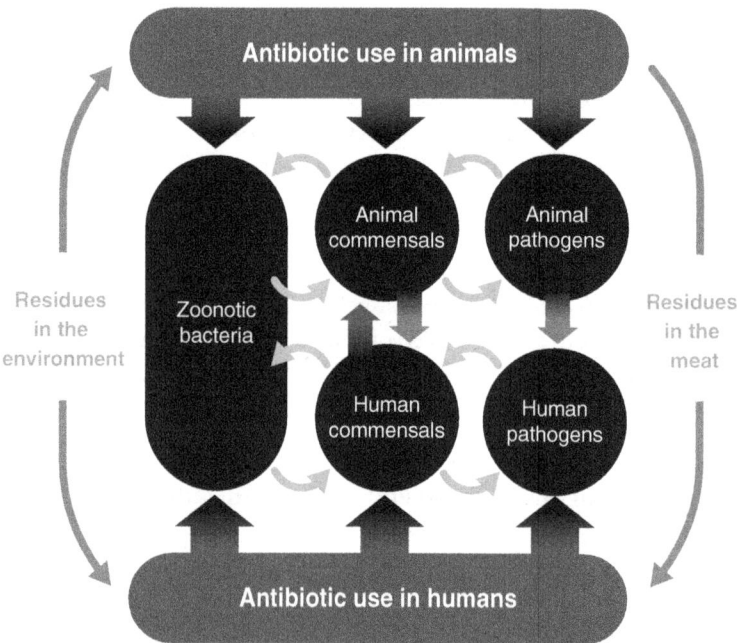

Figure 5.2 Different routes for exchange of resistant bacteria or genes from animals to humans and vice versa

Source: Authors' compilation.

genes) may be transferred between animals and humans through direct contact. It has long been established that farmers and farm workers have higher levels of resistant bacteria than people who do not live close to livestock. Similarly, hospitals can act as a hot spot for AMR, exposing both humans and animals that live nearby.

Companion animals should not be overlooked in the debate concerning transmission of resistance because of their close contact with humans. It is not surprising that an increasing amount of scientific literature describes resistance transmission between companion animals and humans (Pomba et al., 2017).

Resistant traits can also reach humans through the consumption of food that contains resistant bacteria or genes. The most obvious form of foodborne transmission seems to be from the consumption of meat, milk or eggs. Yet if these animal products are heat treated (e.g. cooked or pasteurized), and the required hygienic measures are applied in the kitchen, there should be very little or no transfer of (resistant) bacteria. The consumption of raw products is more risky. Foodborne transmission may also occur as a result of the consumption of vegetables grown in soil fertilized with manure of animal origin, or irrigated with contaminated water. Finally, resistant traits can be spread via waste material contaminating the environment. Water is a particularly efficient and quick route of transmission.

Box 5.5 Animal to human transfer of antibiotic-resistant strains

Livestock-associated methicillin-resistant *Staphylococcus aureus* (LA-MRSA) is frequently found in pigs, and people in contact with pigs, but also in other livestock species (Crombé et al., 2013). Human carriers are typically people in contact with pigs, e.g. farmers, farm workers, veterinarians and slaughterhouse staff. Another example is the identification of shared reservoirs of extended spectrum beta-lactamase Enterobacteriaceae genes between humans and animals (Madec et al., 2017). Therefore, exposure to animals is regarded as a risk factor and indirect transmission is not unlikely.

Other interventions to reduce antimicrobial use in food animal production

There are numerous other ways to reduce AMU. Some are generically oriented towards disease prevention, such as the improvement of water quality and biosecurity. Others are designed to address specific pathogens, such as specific-pathogen-free (SPF) programmes. Some concern overall management or culture (e.g. benchmarking of veterinarians to raise awareness to differences in culture and traditions when prescribing antimicrobial agents), while others are of a biological nature, such as probiotics or prebiotics, or the use of genetically enhanced breeds that are less susceptible to certain diseases.

In a recent study (Postma et al., 2015), European veterinarian practitioners active in pig production were asked what they consider to be the most valid alternatives for AMU in pig production, taking into account expected effectiveness, feasibility and return on investment of the measures. Results indicated that practitioners believe the most promising alternatives to AMU are, in order of priority: improved biosecurity, increased and improved vaccination, use of zinc, improved feed quality and improved diagnostics.

In recent years, several studies have found that improved biosecurity may result in reduced AMU, without jeopardizing production results. In a study in breeder–finisher pig herds in Belgium, it was found that herds with higher internal biosecurity scores had lower antimicrobial treatment incidences, suggesting that improved biosecurity might help in reducing the amount of antimicrobials used (Laanen et al., 2013). In a French study in breeder–finisher herds, biosecurity measures such as disinfection of the loading area, gilt quarantine and adaptation, farm structure/working lines and all in/all out practices were found to be significantly associated with lower AMU (Lannou et al., 2012). In a recent study in four European countries, it was shown that a higher weaning age, a week system of five weeks or more and the external biosecurity level were significantly associated with a lower antimicrobial treatment incidence (Postma et al., 2016). This finding was confirmed in a study of the profile of top farmers. In this study, the level of internal biosecurity was positively associated with a better control of infectious diseases and a lower need for antimicrobials (Collineau et al., 2017a).

In Denmark, measures were implemented by farmers and their veterinarians that managed to reduce their annual antimicrobial consumption

---

Box 5.6 Importance of cost–benefit of biosecurity reporting

Farmers often perceive improvements in biosecurity as difficult to achieve and not cost-effective, mainly because they lack information on their associated costs and, especially, revenues (Fraser et al., 2010; Laanen et al., 2014). One study made an inventory of the application of biosecurity measures in 77 breeder–finisher herds in France. It showed that the difference in standardized profit margins between farms with high biosecurity compared with those with lower levels of biosecurity were estimated at around €200 per sow per year (Corrégé et al., 2012).

---

by 10% or more following the introduction of the "Yellow Card system". It was reported, among other parameters, that cleaning procedures, adequate action regarding diseased animals (e.g. an earlier decision to euthanize) and all in/all out were mentioned by farmers and veterinarians as means to reduce AMU (Dupont et al., 2017). Another study concluded that improved biosecurity, especially the presence of a hygiene lock, and pest control by a professional, were related to lower probabilities of farms being infected with extended spectrum beta-lactamase *E. coli* (Dohmen et al., 2017).

An intervention study in Belgium found that improving pig herd management and biosecurity status, in combination with antimicrobial stewardship, helped reduce AMU from birth till slaughter by 52%, and in sows by 32% (Postma et al., 2017). In this study, the management and biosecurity interventions were generally relatively simple for farmers to implement. They included changing the working habits and routines of the farmer (e.g. changing of needles, hand and personal hygiene, and analysis of water quality). Interventions incurring higher costs and/or more pronounced changes, such as introducing a new hygiene lock to change clothes/boots and wash hands, were implemented less frequently. A key recommendation was for a good and early registration of disease symptoms in order to be able to take proper and timely control measures (e.g. biosecurity, vaccination and climate change), and to create awareness of the importance of the principle that "prevention is better than cure".

An important success factor from the above study was the order of action: "Check, Improve and Reduce". It suggested that herd counselling should always start with a thorough evaluation of herd management, biosecurity and health status, followed by tailored advice with specific suggestions for improvement. In this process it was important that an adviser/coach helped the farmer by explaining what he/she could improve, and what the risk would be when certain practices were not performed correctly. In addition, follow-up and feedback on the agreed and executed improvements is of high importance to retain levels of motivation. Only after implementation of these improvements, may changes and reductions in AMU be proposed. Using this approach, farmers can keep control over the health situation and are less reluctant to change certain antimicrobial treatment procedures. An economic evaluation based on the results of the study has shown that, including labour costs of all persons involved (including the coach, veterinarian and farmer), the participating herds achieved an average financial gain or overall benefit of €2.67 per finisher pig per year from partaking in this "team effort" approach (Rojo-Gimeno et al., 2016). In a comparable study performed in four European Union countries, an economic evaluation of suggested interventions in, among others, biosecurity resulted in a median change in net farm profits among Belgian and French farms estimated at €4.46 and €1.23 per sow per year, respectively (Collineau, Rojo-Gimeno & Léger, 2017b). A comparable type of intervention study performed in Belgium on 15 broiler farms, based on improved biosecurity, recorded an average reduction of 29% in AMU (Gelaude et al., 2014).

In other animal species, studies about the association between biosecurity and AMU are scarce. However, the results obtained in pig production may be applicable to other farm animals. Experiences from the introduction of high biosecurity standards in Swedish broiler production to reduce the risk of Salmonella have demonstrated the close association between improved biosecurity and reduced AMU. Improvements in the level of biosecurity should at least be at the basis of any effort to reduce AMU at herd or flock level.

Besides the above described effects of improved biosecurity, other methods such as improved vaccination, use of feed and water additives and an improved feed regimen are also available. For example, essential oils, prebiotics (feed ingredients with beneficial effects on the gut microbiota) and probiotics (microorganisms with beneficial health effects) have been proposed for managing post-weaning diarrhoea in piglets (Gresse et al., 2017) and various probiotics have been developed to control necrotic enteritis in broiler chickens (Caly et al., 2015). Feed additives such as probiotics, prebiotics, organic acids and hyper-immune egg yolk antibodies have also been used to enhance the growth of broiler chickens (Gadde et al., 2017). However, despite a wide range of new potential alternatives to antibiotics, including vaccines, other immunomodulators, bacteriophages, lysins, hydrolases, antimicrobial peptides, plant extracts, quorum sensing inhibitors, biofilm inhibitors, bacterial virulence inhibitors, enzymes, pre-, pro- and synbiotics it has been concluded that antibiotic resistance and tolerance in bacteria are natural evolutionary consequences and in the foreseeable future prudent use of antibiotics is the best and fastest way to limit the growth of AMR (Cheng et al., 2014; Sang & Blecha, 2015).

Conclusions

The global use of antibiotics in animal production has been excessive and contributed to the selection of antibiotic resistance affecting both human and animal health. The realization that even low doses of anti-microbials, such as are used for growth promotion in animals and seen in agricultural waste, exerts a selective pressure for increasing AMR among bacteria in the environment, animals and humans has sparked a range of global activities to counteract these effects.

In recent years, however, huge progress has been made in the field of improved animal management. In addition, new tools for disease preven-tion and control are being developed. To ensure a global implementation of these tools for better animal management and more prudent use of antibiotics in animal production, significant efforts will be required in several areas. These include increasing awareness of the risks associated with AMR, improving training and education on the use of antibiotics, enhancing external and internal biosecurity measures, and improving the overall husbandry and management practices on many animal farms.

Implementation of these measures indicates already that the use of antimicrobials in animal production can be substantially reduced in the future without a negative impact on production and animal health and welfare. This reduction will also result in the checking, and eventually even reversal, of resistance selection which will have further benefits for animal health and production as well as human health, global food safety and food security.

References

Aarestrup FM, Jensen VF, Emborg HD et al. (2010). Changes in the use of antimicrobials and the effects on productivity of swine farms in Denmark. Am J Vet Res. 71:726–733.

Adeyi OO, Baris E, Jonas OB et al. (2017). Drug-resistant infections: a threat to our economic future (Vol.2): final report. Washington, DC: World Bank Group. (http://documents.worldbank.org/curated/en/323311493396993758/final-report, accessed 06 September 2018).

bioMérieux (2016). Bacteria and the environment. Marcy-l'Étoile: bioMérieux. (https://www.antimicrobial-resistance.biomerieux.com/ popup/bacteria-and-the-environment/, accessed 06 September 2018).

Bos ME, Taverne FJ, van Geijlswijk IM et al. (2013). Consumption of antimicrobials in pigs, veal calves, and broilers in the Netherlands. Quantitative Results of Nationwide Collection of data in 2011. PLoS One. 8(10):e77525.

Broom LJ (2017). The sub-inhibitory theory for antibiotic growth promoters. Poult Sci. 96:3104–3108.

Burow E, Simoneit C, Tenhagen BA et al. (2014). Oral antimicrobials increase antimicrobial resistance in porcine E. coli – a systematic review. Prev Vet Med. 113(4):364–375.

Cabello FC (2006). Heavy use of prophylactic antibiotics in aquaculture: a growing problem for human and animal health and for the environment. Environ Microbiol. 8:1137–1144.

Callens B, Cargnel M, Sarrazin S et al. (2017). Associations between a decreased veterinary antimicrobial use and resistance in commensal Escherichia coli from Belgian livestock species (2011–2015). Prevent Vet Med. 157:50–58.

Caly DL, D'Inca R, Auclair E et al. (2015). Alternatives to antibiotics to prevent necrotic enteritis in broiler chickens: a microbiologist's perspective. Front Microbiol. 6:1336.

Chantziaras I, Boyen F, Callens B et al. (2014). Correlation between veterinary antimicrobial use and antimicrobial resistance in foodproducing animals: a report on seven countries. J Antimicrob Chemother. 69(3):827–834.

Charbonneau P, Parienti JJ, Thibon P et al. (2006). Fluoroquinolone use and methicillin-resistant Staphylococcus aureus isolation rates in hospitalized patients: a quasi experimental study. Clin Infect Dis. 42(6):778–784.

Cheng G, Hao H, Xie S et al. (2014). Antibiotic alternatives: the substitution of antibiotics in animal husbandry? Front Microbiol. 5:217.

Collineau L, Belloc C, Stärk KD et al. (2016). Guidance on the selection of appropriate indicators for quantification of antimicrobial usage in humans and animals. Zoonoses Public Health. 64(3):165–184.

Collineau L, Backhans A, Dewulf J et al. (2017a). Profile of pig farms combining high performance and low antimicrobial usage within four European countries. Vet Rec. 18:657.

Collineau L, Rojo-Gimeno C, Léger A. (2017b). Herd-specific interventions to reduce antimicrobial usage in pig production without jeopardising technical and economic performance. Prevent Vet Med. 144:167– 178.

Corrégé I, Fourchon P, Le Brun T et al. (2012). Biosécurité et hygiène en élevage de porcs: état des lieux et impact sur les performances technico-économiques. Journées Recherche Porcine. 44:101–102.

Costelloe C, Metcalfe C, Lovering A et al. (2010). Effect of antibiotic prescribing in primary care on antimicrobial resistance in individual patients: systematic review and meta-analysis. BMJ. 340:c2096.

Crombé F, Argudín MA, Vanderhaeghen W et al. (2013). Transmission dynamics of methicillin-resistant Staphylococcus aureus in pigs. Front Microbiol. 4:57.

Dohmen W, Dorado-García A, Bonten MJ et al. (2017). Risk factors for ESBL-producing Escherichia coli on pig farms: A longitudinal study in the context of reduced use of antimicrobials. PLoS One. 12:e0174094.

Done HY, Venkatesan AK, Halden RU (2015). Does the recent growth of aquaculture create antibiotic resistance threats different from those associated with land animal production in agriculture? AAPS. 17(3):513.

Dorado-García A, Mevius DJ, Jacobs JJ et al. (2016). Assessment of antimicrobial resistance in livestock during the course of a nationwide antimicrobial use reduction in the Netherlands. J Antimicrob Chemother. 71(12):3607–3619.

Dupont N, Diness LH, Fertner M et al. (2017). Antimicrobial reduction measures applied in Danish pig herds following the introduction of the "Yellow Card" antimicrobial scheme. Prevent Vet Med. 138:9–16.

European Centre for Disease Prevention and Control (ECDC), European Food Safety Authority (EFSA), European Medicines Agency (EMA) (2017). ECDC/EFSA/EMA second joint report on the integrated analysis of the consumption of antimicrobial agents and occurrence of antimicrobial resistance in bacteria from humans and foodproducing animals: Joint Interagency Antimicrobial Consumption and Resistance Analysis (JIACRA) Report. EFSA. 15(7):4872.

European Food Safety Authority (EFSA), European Centre for Disease Prevention and Control (ECDC) (2018). The European Union summary report on antimicrobial resistance in zoonotic and indicator bacteria from humans, animals and food in 2016. EFSA. 16(2):5182.

European Medicines Agency (EMA) (n.d.). European Surveillance of Veterinary Antimicrobial Consumption (ESVAC): ESVAC strategy 2016–2020. (http:// www.ema.europa.eu/ema/index.jsp?curl=pages/regulation/document_ listing/document_listing_000302.jsp. accessed 06 September 2018).

Fraser RW, Williams NT, Powell LF, Cook AJ (2010). Reducing Campylobacter and Salmonella infection: Two studies of the economic cost and attitude to adoption of on-farm biosecurity measures. Zoonoses Public Health. 57:e109-e115.

Gadde U, Kim WH, Oh ST et al. (2017). Alternatives to antibiotics for maximizing growth performance and feed efficiency in poultry: a review. Animal Health Res Rev. 18:26–45.

Gelaude P, Schlepers M, Verlinden M et al. (2014). Biocheck.UGent: A quantitative tool to measure biosecurity at broiler farms and the relationship with technical performances and antimicrobial use. Poult Sci. 93:2740–2751.

Grave K, Kaldhusdal MC, Kruse H et al. (2004). What has happened in Norway after the ban of avoparcin? Consumption of antimicrobials by poultry. Prevent Vet Med. 62(1):59–72.

Grave K, Jensen VF, Odensvik K et al. (2006). Usage of veterinary therapeutic antimicrobials in Denmark, Norway and Sweden following the termination of antimicrobial growth promoter use. Prevent Vet Med. 75(1–2):123–132.

Gresse R, Chaucheyras-Durand F et al. (2017). Gut microbiota dysbiosis in postweaning piglets: understanding the keys to health. Trends Microbiol. 25(10):851–873.

Gulla S, Duodu S, Nilsen A et al. (2016). Aeromonas salmonicida infection levels in pre- and post-stocked cleaner fish assessed by culture and an amended qPCR assay. J Fish Dis. 39(7):867–877.

Hammerum AM, Larsen J, Andersen VD et al. (2014). Characterization of extended-spectrum β-lactamase (ESBL)-producing Escherichia coli obtained from Danish pigs, pig farmers and their families from farms with high or no consumption of third- or fourth-generation cephalosporins. J Antimicrob Chemother. 69(10):2650–2657.

Krishnasamy V, Otte J, Silbergeld E (2015). Antimicrobial use in Chinese swine and broiler poultry production. Antimicrob Resist Infect Control. 4:17.

Kristensen E, Jakobsen EB. (2011). Danish dairy farmers' perception of biosecurity. Prev Vet Med. 99(2–4):122–129.

Laanen M, Persoons D, Ribbens S et al. (2013). Relationship between biosecurity and production/antimicrobial treatment characteristics in pig herds. Vet J. 198:508–512.

Laanen M, Maes D, Hendriksen C et al. (2014). Pig, cattle and poultry farmers with a known interest in research have comparable perspectives on disease prevention and on-farm biosecurity. Prevent Vet Med. 115:1–9.

Landers TF, Cohen B, Wittum TE et al. (2012). A review of antibiotic use in food animals: Perspective, policy and potential. Pub Health Rep. 127:4–22.

Lannou J et al. (2012). Antibiotiques en élevage porcin: modalites d'usage et relation avec les pratiques d'elevage. Toulouse: Association Française de Médecine Vétérinaire Porcine.

Laxminarayan RT, Van Boeckel T, Teillant A (2015). The economics costs of withdrawing antimicrobial growth promoters from the livestock sector. OECD Food, Agriculture and Fisheries Papers. Paris: OECD Publishing. (https://www.oecd-ilibrary.org/agriculture-and-food/the-economic-costs-of-withdrawing-anti-microbial-use-in-the-livestock-sector_5js64kst5wvl-en, accessed 06 September 2018).

Liu Y-Y, Wang Y, Walsh TR et al. (2016). Emergence of plasmid-mediated colistin resistance mechanism MCR-1 in animals and human beings in China: a microbiological and molecular biological study. Lancet Infect Dis. 16:161–168.

Madec J-Y, Haenni M, Nordmann P (2017). Extended-spectrum β-lactamase/ Amp-C and carbapenemase-producing Enterobacteriaceae in animals: a threat for humans? Clin Microbiol Infect. 23(11):826–833.

OECD (2017). Producer incentives in livestock disease management: a synthesis of conceptual and empirical studies. Paris: OECD Publishing. (http://www.oecd.org/publications/producer-incentives-in-livestock-disease-management-9789264279483-en.htm, accessed 06 September 2018).

Pomba C, Rantala M, Greko C et al. (2017). Public health risk of antimicrobial resistance transfer from companion animals. J Antimicrob Chemother. 72:957–968.

Postma M, Stärk KD, Sjölund M et al. (2015). Alternatives to the use of antimicrobial agents in pig production: A multi-country expert ranking of perceived effectiveness, feasibility and return on investment. Prevent Vet Med. 118:457–466.

Postma M, Backhans A, Collineau L et al. (2016). Evaluation of the relationship between the biosecurity status, production parameters, herd characteristics and antimicrobial usage in farrow-to-finish pig production in four EU countries. Porcine Health Manag. 2:9. doi: 10.1186/s40813-016-0028-z.

Postma M, Vanderhaeghen W, Sarrazin S et al. (2017). Reducing antimicrobial usage in pig production without jeopardizing production parameters. Zoonoses Public Health. 64:63–74.

Racicot M, Venne D, Durivage A et al. (2012). Evaluation of the relationship between personality traits, experience, education and biosecurity compliance on poultry farms in Québec, Canada. Prev Vet Med. 103(2–3):201–207.

Rhouma M, Fairbrother JM, Beaudry F et al. (2017). Post weaning diarrhea in pigs: risk factors and non-colistin-based control strategies. Acta Veterinaria Scandinavica. 59:31.

Rojo-Gimeno C, Postma M, Dewulf J et al. (2016). Farm-economic analysis of reducing antimicrobial use whilst adopting good management strategies on farrow-to-finish pig farms. Prevent Vet Med. 129:74–87.

Rushton J (2013). An overview analysis of costs and benefits of government control policy options. Paris: OECD Publishing. (https://www.oecd.org/tad/agricultural-policies/AN%20OVERVIEW%20OF%20ANALYSIS%20OF%20COSTS%20AND%20BENEFITS.pdf, accessed 06 September 2018).

Rushton J (2015). Antimicrobial use in animals: How to assess the tradeoffs. Zoonoses Public Health. 62:10–21.

Rushton J, Pinto Ferreira J (2014). Antimicrobial resistance: The use of antimicrobials in the livestock sector. OECD Food, Agriculture and Fisheries Papers. Paris: OECD Publishing. (https://www.oecd-ilibrary.org/agriculture-and-food/antimicrobial-resistance_5jxvl3dwk3f0-en, accessed 06 September 2018).

Sang Y, Blecha F (2015). Alternatives to antibiotics in animal agriculture: An ecoimmunological view. Pathogens. 4:1–19.

Simoneit C, Burow E, Tenhagen BA (2015). Oral administration of antimicrobials increase antimicrobial resistance in E. coli from chicken – a systematic review. Prev Vet Med. 118(1):1–7.

Sneeringer SJ, MacDonald JM, Key N et al. (2015). Economic research report No 200: Economics of antibiotic use in U.S. livestock production. Washington, DC: United States Department of Agriculture. (https://www.ers.usda.gov/publications/pub-details/?pubid=45488, accessed 06 September 2018).

Speksnijder DC, Mevius DJ, Bruschke CJ et al. (2015). Reduction of veterinary antimicrobial use in the Netherlands. The Dutch success model. Zoonoses Public Health. 62:79–87.

Sun L, Klein EY, Laxminarayan R (2012). Seasonality and temporal correlation between community antibiotic use and resistance in the United States. Clin Infect Dis. 55(5):687–694.

Swedish Ministry of Agriculture (1997). Report from the commission on antimicrobial feed additives. Government Official Reports. Stockholm: Swedish Ministry of Agriculture. (http://www.regeringskansliet.se/rattsdokument/statens-offentliga-utredningar/1997/01/sou-1997132/, accessed 06 September 2018).

Wall BA, Mateus A, Marshall L et al. (2016). Drivers, dynamics and epidemiology of antimicrobial resistance in animal production. Rome: Food and Agriculture Organization of the United Nations. (http://www.fao.org/documents/card/en/c/d5f6d40d-ef08-4fcc-866b-5e5a92a12dbf/, accessed 06 September 2018).

World Health Organization (2015). Global action plan on antimicrobial resistance. Geneva: World Health Organization. (http://www.who.int/antimicrobial-resistance/global-action-plan/en/, accessed 06 September 2018).

World Organisation for Animal Health (OIE) (2016). OIE Strategy tackles the threat of Antimicrobial Resistance (AMR) in animals. Paris: World Organisation for Animal Health. (http://www.oie.int/for-the-media/press-releases/detail/article/oie-strategy-tackles-the-threat-ofantimicrobial-resistance-amr-in-animals/, accessed 06 September 2018).

World Organisation for Animal Health (OIE) (2018). Antimicrobial resistance: OIE activities. Paris: World Organisation for Animal Health. (http://www.oie.int/scientific-expertise/veterinary-products/antimicrobials/, accessed 06 September 2018).

6 Fostering R&D of novel antibiotics and other technologies to prevent and treat infection

MATTHEW RENWICK, ELIAS MOSSIALOS

Introduction

In the past, developing new antibiotics appeared to be the easiest solution to overcome resistant pathogens. As certain antibiotics became less effective against evolving bacteria, treatment for infections could be supplemented or replaced by newer generations of the same antibiotic or by a new, more effective class of antibiotic. The world saw a boom in new antibiotics and classes between 1940 and 1990 as pharmaceutical companies leveraged scientific breakthroughs and were rewarded with high-value patents (Pew Charitable Trusts, 2016; Silver, 2011). However, due to a combination of financial, regulatory, and scientific barriers to continued development of new antibiotics, the focus of research and development (R&D) shifted away to other therapeutic areas (Renwick & Mossialos, 2018). In 1990, there were 18 major pharmaceutical companies active in antibiotic R&D, but today there are only eight (Access to Medicine Foundation, 2018; Butler, Blaskovich & Cooper, 2013). Since then, the number of new antibiotics marketed each decade has significantly decreased and no novel classes of antibiotics with distinct chemical structures have been developed (Pew Charitable Trusts, 2016) (Figure 6.1). This void in discovery and development has meant that the antibiotic pipeline is frighteningly thin relative to the unrelenting advance of antibiotic resistance.

The global community is beginning to accept the severity of antibiotic resistance and is scrambling to make up for lost time in antibiotic R&D. Promisingly, numerous major international and national initiatives have been started in recent years to help fund, coordinate, and incentivize antibiotic R&D programmes (Simpkin et al., 2017). With the recent flurry of action it is important, however, to take stock and assess the current state of the global market for antibiotics and antibiotic innovation, in order to identify any necessary policy adjustments. To this end, this chapter aims to identify progress and assess the challenges in

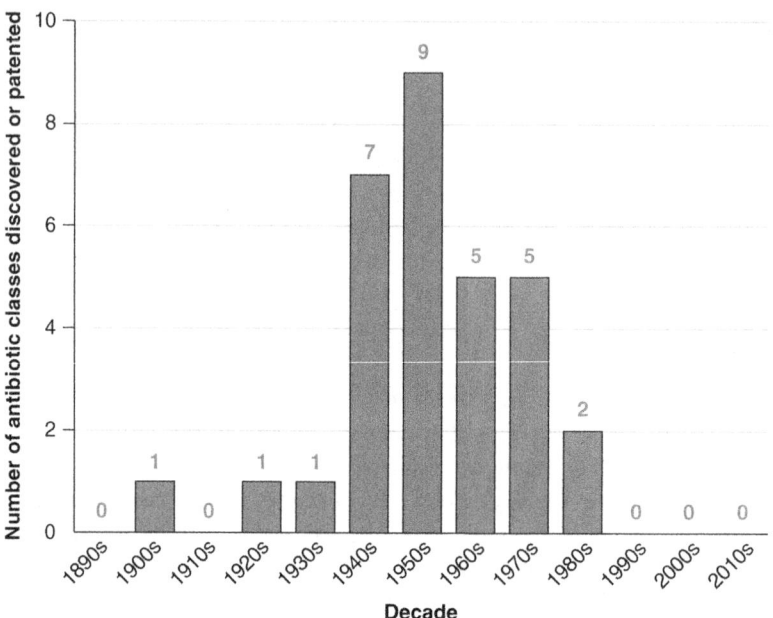

Figure 6.1 Number of new classes of antibiotic discovered or patented each decade

Source: Pew Charitable Trusts, 2016. Originally adapted from Silver, 2011.

fostering antibiotic R&D, as well as highlighting some key policy gaps that must be addressed.

The antibiotic pipeline

Although the antibiotic pipeline is improving, it is not nearly robust enough to match clinical need and respond to the rising rates of resistance in deadly pathogens. In early 2017, the World Health Organization (WHO) (2017a) published a priority pathogens list (PPL), which outlines the antibiotic-resistant bacteria that pose the greatest threat to global public health (Table 6.1). This list aims to guide antibiotic R&D based on medical need as opposed to the economic factors that have traditionally directed antibiotic investment. At the top of this list, categorized as "critical", are the Gram-negative, carbapenem-resistant strains of *Acinetobacter baumannii, Pseudomonas aeruginosa,* and the Enterobacteriaceae family.

Table 6.1 *WHO Priority Pathogens List (PPL): Global priority list of antibiotic-resistant bacteria to guide research, discovery, and development of new antibiotics*

Priority level	Pathogens
Critical	*Acinetobacter baumannii*, carbapenem-resistant *Pseudomonas aeruginosa*, carbapenem-resistant *Enterobacteriaceae*, carbapenem-resistant & third-generation cephalosporin-resistant
High	*Enterococcus faecium*, vancomycin-resistant *Staphylococcus aureus*, methicillin-resistant, vancomycin intermediate and resistant *Helicobacter pylori*, clarithromycin-resistant *Campylobacter*, fluoroquinolone-resistant *Salmonella* spp., fluoroquinolone-resistant *Neisseria gonorrhoeae*, third-generation cephalosporin-resistant, fluoroquinolone-resistant
Medium	*Streptococcus pneumoniae*, penicillin-non-susceptible *Haemophilus influenzae*, ampicillin-resistant *Shigella* spp., fluoroquinolone-resistant

Source: World Health Organization, 2017a.

In 2013, the US Centers for Disease Control and Prevention (CDC) published a US-focused urgent threats list for antibiotic resistance, which highlighted many of the same pathogens (US CDC, 2013).

In September 2017, the WHO published an in-depth analysis of the global development pipeline for antibacterial agents (World Health Organization, 2017b). The report shows how 32 antibiotic therapies that are active or possibly active against a PPL pathogen are the subject of clinical trials: 14 in phase I clinical trials, 8 in phase II, and 10 in phase III (Figure 6.2). Based on optimistic clinical trial attrition rates, the report estimates that the entire pipeline could be expected to yield 10 new approvals. As with most drug developments, the R&D and market approval process is lengthy. The phase III antibiotics are three to five years from potentially reaching the market. However, the phase I and II antibiotics have development timelines of at least five to 10 years and successful progression to marketing approval is far from certain. Antibiotics in phase I clinical trials have only a 14% likelihood

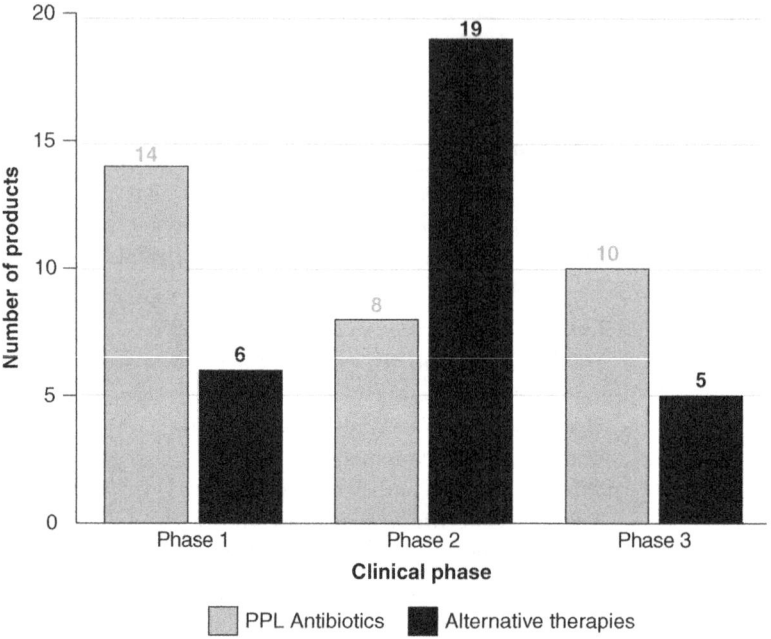

Figure 6.2 The number of antibiotics in clinical development possibly active against WHO PPL pathogens (2017) and the number of alternative therapies to antibiotics in clinical development (2017)

Note: PPL: Priority Pathogens List.

Source: World Health Organization, 2017b; Pew Charitable Trusts 2017.

of reaching the market. This means that of the 10 phase I antibiotics targeting resistant Gram-negative bacteria only one or two will succeed.

A compounding problem is that most of the pipeline drugs are redevelopments of classic antibiotic compounds or are combination therapies of existing antibiotic molecules. These types of less original antibiotics are at higher risk of quickly losing effectiveness in clinical practice because of cross-resistance. According to the WHO pipeline analysis, there are eight products in development that offer innovation in terms of having at least one of the following criteria: (1) absence of cross-resistance to existing antibiotics, (2) new chemical class, (3) new target, or (4) new mechanism of action. There are only two drugs that are truly innovative across all four WHO criteria: one targets *P. aeruginosa* and the other *Staphylococcus aureus*. The WHO analysis concluded that the current antibacterial pipeline is inadequate for the soaring resistance rates (World Health Organization, 2017b). This sentiment was echoed in the Access to Medicine Foundation's 2018 AMR Benchmark Report, which is an independent assessment of key industry players across a spectrum of AMR priorities related to R&D, production and manufacturing, and appropriate access and stewardship (Access to Medicine Foundation, 2018). Of note, Pew Charitable Trusts conducts a concise and useful biannual antibiotic pipeline analysis and the most recent update, as of September 2018, reiterates these general findings (Pew Charitable Trusts, 2018).

A fledgling portfolio of alternative therapies to antibiotics is now emerging and includes vaccines, immune stimulation, bacteriophages, lysins, probiotics, antibodies and various peptides. Initially, these products would likely supplement typical antibiotic regimens as adjunctive or preventive therapies. In March 2017, a Pew Charitable Trusts analysis found that there are 30 nontraditional antibacterial therapies in the development pipeline: six in phase I clinical trials, 19 in phase II, and five in phase III trials (Figure 6.2) (Pew Charitable Trusts, 2017). Of these, nine products are vaccines, nine are antibodies, and the remainder are probiotics, lysins and peptide immunomodulators. However, an earlier review of these alternative treatments estimated that this pipeline will require more than £1.5 billion in sustained investment over 10 years to translate most of these projects into investable ventures and eventually marketable products (Czaplewski et al., 2016). Another ongoing challenge with some alternative therapies, such as bacteriophages, is that there is minimal regulatory precedent for FDA and EMA licensure, making it challenging for developers and regulators to know how to proceed (Cooper, Khan Mirzaei & Nilsson, 2016).

Antibiotic R&D funding

Antibiotic development is funded by a combination of public and private investment and an increasing number of R&D projects are being funded through partnerships (Access to Medicine Foundation, 2018). As most investment data on antibiotic R&D is confidential or unpublished, it is challenging to accurately assess current trends in global funding. However, there is some available data. With regard to public funding, between 2007 and 2013, the European Union (EU) and countries in the Joint Programming Initiative on Antimicrobial Resistance (JPIAMR) invested €1.3 billion across 1 243 research projects on antibacterial resistance (Kelly et al., 2015). Most of this funding supported R&D for antibiotics, alternative therapies, and diagnostics. In 2016 and 2017, the US National Institutes for Health's (NIH) budget for AMR was $420 million and $473 million, respectively – a major portion of this will have been dedicated specifically to antibiotic R&D projects (US National Institutes of Health, 2017). The Biomedical Advanced Research and Development Authority (BARDA), the largest US funding agency for antibiotic R&D outside the NIH, has an annual budget of $192 million to develop therapies treating antibiotic-resistant bacteria (US Department of Health and Human Services, 2017). These US and EU budgets are small in comparison to the public money spent on many noncommunicable diseases such as cancer, which command annual budgets in the billions in the US and EU (Eckhouse & Sullivan, 2006; US National Institutes of Health, 2017). Missing from this picture are the amounts of public funding by other countries with significant investments in pharmaceutical R&D, including Japan, China, India, and the Republic of Korea.

In the private sector, $1.8 billion in global venture capital was invested in antimicrobial R&D between 2004 and 2013 (Thomas & Wessel, 2015). Venture capital investment dropped by 28% between the first and second halves of this 10-year time frame. There are no data available on the investments made by pharmaceutical companies in their own antibiotic projects, but it appears that internal funding of antibiotic R&D is a relatively low priority. For instance, the global number of patent applications related to antibiotic research dropped by 34.8% from 2007 to 2012, which may indicate a decreasing commercial interest in antibiotic R&D (Marks & Clerk, 2015). The WHO's International Clinical Trials Registry Platform shows that there are only 182 active clinical trials focusing on bacterial infections other than tuberculosis, which is much less than 1% of the 67 000 clinical trials on noncommunicable diseases (O'Neill, 2015b). These numbers seem to indicate that the economic case for private investment in antibiotic R&D, both external and internal to pharmaceutical companies, has not improved over the past decade. Public and nongovernmental institutions cannot entirely replace private companies in the development of novel antibiotics. Thus, there is a need for public and philanthropic organizations to increase funding to support private companies in antibiotic R&D and implement non-monetary incentive policies that reduce barriers throughout the antibiotic development value chain.

Barriers to antibiotic R&D

The success rate of moving an antibiotic from basic research to market approval is estimated to be between 1.5% and 3.5%. This process can typically take 15 years (O'Neill, 2015a). The economic, regulatory, and scientific barriers to antibiotic R&D can best be categorized based on the steps of the antibiotic value chain: initial research, preclinical trials, clinical trials, market approval, and, finally, commercialization (Chorzelski et al., 2015). These barriers are important to consider when designing and targeting future incentives to support antibiotic R&D.

The basic and discovery research behind understanding and identifying new molecules for candidate drugs has been scientifically challenging. Bacteria, particularly Gram-negative varieties, have proven highly resilient to recent experimental mechanisms of destruction (Pew Charitable Trusts, 2016). In addition, scientific expertise in this area is currently lacking and is still recovering from the discovery void that began in 1990 (Silver, 2011). The preclinical stage is ominously referred to as the "valley of death" (So et al., 2012). Discovery research has predominantly been tackled by academics funded by the public sector, while clinical trials have been the domain of private pharmaceutical companies, thus leaving a gap in funding and appropriate actors to move from one to the other.

Antibiotic clinical trials are costly, estimated at roughly $130 million to take a drug candidate through Phases I to III. Many drug candidates will be discarded on the way, at a financial loss. The average cost of post-approval follow-on trials can amount to an additional $146 million (O'Neill, 2015a). These costs and uncertainties are often prohibitively high for small and medium enterprises (SMEs) (Renwick, Brogan & Mossialos, 2016). Despite the challenge of economies of scale, SMEs represent approximately 85% of the share of antibiotics in clinical development (Chorzelski et al., 2015). An added practical challenge is that recruiting patients with acute bacterial infections for clinical trials is logistically difficult due to the short treatment windows and lack of rapid point-of-care diagnostic tools to identify potential participants.

Market approval of new antibiotics is necessary for ensuring the drug's quality, safety and efficacy. However, there are procedural differences between drug regulatory agencies in approving antibiotics that make global licensing unduly time-consuming and expensive (Renwick, Simpkin & Mossialos, 2016). These differences relate to patient selection criteria, definitions of clinical end-points, specification of statistical parameters, and rules regarding expedited approvals (Chorzelski et al., 2015).

Finally, the economic reward for commercializing a new antibiotic is minimal or negative relative to other therapeutic areas, such as neurologic or cardiovascular drugs (So et al., 2011). At present, novel antibiotics are not destined to generate significant revenue even with their immense public health value. Potential sales volumes are restricted by short treatment durations and hospital stewardship programmes that limit access. In addition, the large overlap in clinical application of newly patented antibiotics with existing generic alternatives places downward pressure on prices (Renwick, Simpkin & Mossialos, 2016).

Incentivizing antibiotic innovation

A significant amount of research has explored the policy proposals for minimizing these barriers and incentivizing companies to pursue R&D in the antibiotic field. Push and pull incentives are broadly used to classify the two main types of mechanisms for supporting antibiotic R&D (Mossialos et al., 2010). Push incentives reduce the cost of researching and developing new antibiotics. Examples of push incentives include research grants, access to shared resources, and product development partnerships to split R&D costs (Table 6.2). Pull mechanisms increase the potential revenue of a successfully marketed antibiotic. This may be through outcome-based rewards that directly increase revenue such as monetary prizes, reimbursement premiums, advanced market commitments to purchase the drug, and patent buyouts by governments. If large enough, outcome-based pull rewards could replace the traditional revenue stream generated by the sales volumes of a licensed antibiotic. This concept is referred to as "delinkage" since the antibiotic's revenue would be delinked or decoupled from its sales, thus removing the incentive to promote the drug's use (Rex & Outterson, 2016). Alternatively, pull mechanisms may be legal or regulatory, providing incentives such as accelerated procedures for marketing approval or extensions to the patent period. Different push and pull mechanisms have unique advantages and disadvantages and experts generally agree that a combination of both types is necessary to provide effective incentives for R&D. As of 2015, there were 47 different incentives available or proposed for antibiotic developers that ranged from simple push or pull mechanisms to complex hybrid models (Renwick, Brogan & Mossialos, 2016).

Table 6.2 *Push and pull incentives for antibiotic development*

Push incentive strategies

Supporting open access to research	Funding translational research
Grants for scientific personnel	Tax incentives
Direct funding	Refundable tax credits
Conditional grants	Product development partnership

Outcome-based pull incentive strategies

End prize	Research tournament
Milestone prize	Advanced market commitment
Pay-for-performance payments	Strategic Antibiotic Reserve
Patent buyout	Service-availability premium
Payer license	

Lego-regulatory pull incentive strategies

Accelerated assessment and approval	Anti-trust waivers
Market exclusivity extensions	Sui generis rights
Transferable intellectual property rights	Value-based reimbursement
Conservation-based market exclusivity	Targeted approval specifications
Liability protection	Priority review vouchers

Source: Renwick, Simpkin & Mossialos, 2016.

Designing a global incentive package for stimulating antibiotic innovation is a complex task with numerous variables. Policy-makers need a methodology for selecting a complete and realistic set of incentives from the surfeit of candidates. In 2015, the authors of this chapter published a possible framework to help policy-makers with this challenge (Renwick, Brogan & Mossialos, 2016) (Figure 6.3). The framework has three consecutive phases. The first phase involves fashioning a core incentive package targeting the economic criteria necessary for rebalancing the market. This core incentive package must:

1) improve the profitability of developing and commercializing a novel antibiotic;
2) make market participation feasible for SMEs;
3) encourage investment by large pharmaceutical companies;
4) facilitate cooperation across all stakeholders including patients, academics, policy-makers, regulators, and industry.

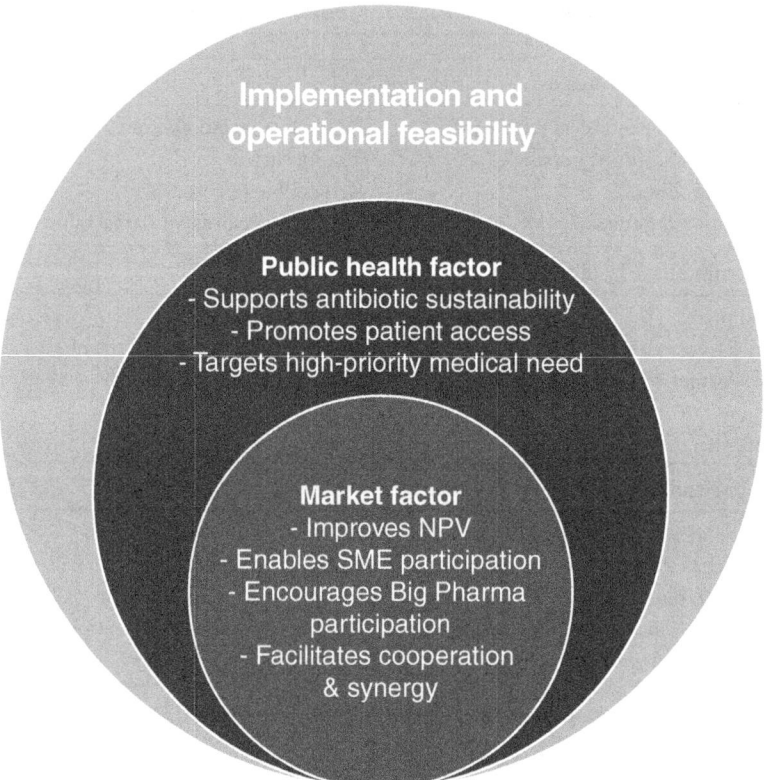

Figure 6.3 Framework for developing a holistic incentive package for antibiotic development

Note: NPV: net present value; SME: Small–medium sized enterprise.

Source: Simpkin et al., 2017.

The second step requires adjusting the core incentive package to address public health goals pertaining to sustainability and patient access to new antibiotics. The final step considers the implementation and operational practicalities that are specific to national context.

Key initiatives incentivizing antibiotic R&D

A 2016 review found that there are 58 active initiatives directly incentivizing the development of antibiotics at global, EU, and national levels, including in the UK, France, Germany, Netherlands, Sweden, the USA, and Canada (Renwick, Simpkin & Mossialos, 2016). These initiatives are programmes that employ one or more push or pull incentive mechanism. An added nine initiatives were identified as offering indirect incentives through economic and policy research or the coordination of strategic actions on AMR. The number of active initiatives in this field continues to rise and several programmes have been initiated since this review was conducted. The following section describes the main antibiotic R&D initiatives at multilateral and EU levels, as well as key initiatives from the USA and the UK, who are leaders at the national level in this field.

Multilateral initiatives

The international community has come together to create several multilateral initiatives including the Joint Programming Initiative on Antimicrobial Resistance (JPIAMR), the Global Antibiotic Research and Development Partnership (GARDP), the Combating Antibiotic Resistant Bacteria Biopharmaceutical Accelerator (CARB-X), the European and Developing Countries Clinical Trial Partnership (EDCTP), and the Global Antimicrobial Resistance Innovation Fund (GAMRIF).

The JPIAMR is comprised of 26 countries with the purpose of coordinating the national funding of its members towards specific AMR research projects, some of which target issues pertaining to R&D. To date, the initiative has funded six joint research calls using a budget of €67 million (Joint Programming Initiative on Antimicrobial Resistance, 2018). Their funding is push-based and is almost exclusively directed towards academic research of basic and preclinical science (Renwick, Simpkin & Mossialos, 2016).

GARDP is a non-profit initiative that is jointly managed by the Drugs for Neglected Diseases Initiative and the WHO. The GARDP's strategic objective is to develop treatments that target the WHO priority pathogens, address diseases and syndromes with the greatest medical need, and help neglected patient populations. As of September 2017, GARDP had secured over €56 million in seed funding of their ultimate funding goal of €270 million (Global Antibiotic Research and Development Partnership, 2017). This initiative is unique in its offering of both push and pull incentives to antibiotic R&D projects, with the possibility of delinking antibiotics that are developed and marketed with the help of GARDP (Renwick, Simpkin & Mossialos, 2016; Global Antibiotic Research and Development Partnership, 2017).

CARB-X is a transatlantic public–private partnership that aims to accelerate basic science and preclinical R&D for a large portfolio of antibiotics, rapid diagnostic tools, and other antimicrobial products. CARB-X has a $505 million investment plan until 2021 with funding support from BARDA, the US National Institute of Allergy and Infectious Diseases (NIAID), the UK's Wellcome Trust, GAMRIF, and the Bill & Melinda Gates Foundation (CARB-X, 2018). Several private partners provide expert scientific and commercial support to their projects. While leadership has initially been in the USA and the UK, this partnership has the capacity to accept additional international partners. As of 2018, CARB-X has accepted 33 projects with a total funding of $91.1 million (CARB-X, 2018). CARB-X projects receive initial push funding with scientific and business guidance. Successful projects can unlock additional funding by reaching certain development milestones. For instance, the initial portfolio of companies and research teams has the potential to access $96.5 million in milestone-based financing (CARB-X, 2018). The CARB-X portfolio will focus on R&D of therapies targeting the pathogens on the CDC's AMR threat list or WHO PPL.

The EDCTP is a public–private partnership that brings together European countries, sub-Saharan African countries, and the pharmaceutical industry to facilitate clinical trials on therapies treating poverty-related communicable diseases that bear the greatest health burden in sub-Saharan Africa. These infections include HIV/AIDS, tuberculosis, malaria, and many neglected infectious diseases. The EDCTP is now in its second decade of operation (2014–2024). From 2014 to 2016, it funded five clinical projects on neglected infectious diseases with a budget of €5.34 million, most of which was directed towards developing new diagnostics (European and Developing Countries Clinical Trials Partnership, 2017).

Lastly, GAMRIF is a new international R&D investment fund spearheaded by the UK Government following recommendations from the UK Review on AMR. The fund supports public and private AMR research ventures that have struggled to attain funding through traditional financing avenues (Simpkin et al., 2017). The UK Government has committed £50 million from 2017 to 2021 to GAMRIF (UK Government, 2016b). As part of a new UK–China research partnership, the Chinese government along with support of private businesses have added a further £10 million to the fund (UK Government, 2016a). In 2018, GAMRIF contributed £20 million to CARB-X for developing vaccines and antibiotic alternatives to treat resistant bacterial infections, as well as £1 million to GARDP for development of an antibiotic for drug-resistant gonorrhoea (UK Government, 2018).

EU initiatives

The EU has been a leader in initiating policy action to revitalize the antibiotic market. The key EU initiatives fostering antibiotic R&D are the European Commission's Directorate-General for Research and Innovation (DG-RTD), the Innovative Medicine's Initiative (IMI), and the InnovFin Infectious Diseases Facility (InnovFin ID).

The DG-RTD partially administers and funds two of the largest antibiotic R&D funding programmes, the EDCTP and the IMI. Beyond these specific programmes, it provides funding support to numerous smaller R&D projects. Between 2007 and 2013, the DG-RTD gave €235.6 million in direct funding for European antibiotics and diagnostics R&D projects, which were separate from the IMI and EDCTP (Kelly et al., 2015). This funding is primarily push-based via direct project funding, research grants, and fellowships. It specifically offers funding opportunities to SMEs undertaking antibiotic R&D through the SME Instrument (Renwick, Simpkin & Mossialos, 2016). In addition, the DG-RTD has created the Horizon 2020 Better Use of Antibiotics Prize, a €1 million market entry reward for creating a rapid point-of-care diagnostic tool for suspected upper respiratory infections (European Commission, 2015).

The IMI is a public–private partnership between the EU and the European Federation of Pharmaceutical Industries and Associations (EFPIA). It has a subsidiary public–private partnership called the New Drugs for Bad Bugs (ND4BB) programme, which is dedicated to the discovery and development of novel antibiotics for humans. Funding for the ND4BB programme is split between the EU and EFPIA and totals €700 million (Innovative Medicines Initiative, n.d.). There are seven core projects, which offer push-based support to most aspects of the antibiotic value chain: TRANSLOCATION and ENABLE assist early drug discovery, COMBACTE supports clinical development of antibiotics for Gram-positive bacteria, COMBACTE-CARE, COMBACTE-MAGNET and iABC facilitate clinical development of antibiotics for Gram-negative bacteria, and DRIVE-AB explores economic solutions to stimulating antibiotic R&D in a sustainable manner. DRIVE-AB's final report with recommendations was published in early 2018 (DRIVE-AB, 2018).

InnovFin ID is a financial risk-sharing programme for ventures in the clinical development phase for a novel drug, vaccine, or diagnostic device that tackles an infectious disease. It is jointly governed by the European Commission and the European Investment Bank (EIB). InnovFin ID offers loans ranging from €7.5 million to €75 million, which are only repaid if the project successfully results in a marketable product. These loans are available to non-profit and for-profit ventures alike (European Investment Bank, n.d.). In autumn 2017, Da Volterra, a small biopharmaceutical firm, entered a €20 million financial agreement with the EIB to support clinical development of their antibiotic portfolio (European Investment Bank, 2017).

US initiatives

There are two US governmental bodies that run programmes to incentivize antibiotic R&D. The first is the NIAID, a research institute within the NIH responsible for conducting basic science and applied research in the field of infectious, immunological, and allergic diseases. The NIAID's AMR portfolio runs from basic science projects to clinical trials for antibiotic therapies, rapid point-of-care diagnostic tools, and vaccines for resistant bacterial infections. The NIH-wide funding for combating AMR in 2017 is $473 million (US National Institutes of Health, 2017). The NIAID supports the Antibiotic Resistant Leadership Group, which is an academic team that prioritizes, designs, and executes clinical research on antibiotic resistance. Additionally, the NIAID is a partner of CARB-X. NIAID's antibiotic R&D incentivization is primarily through direct project funding and research grants (Renwick, Simpkin & Mossialos, 2016).

The second is BARDA, which is an organization within the Office of the Assistant Secretary for Preparedness and Response in the Department of Health and Human Services. BARDA is responsible for facilitating R&D and public purchasing of critical drugs, vaccines, and diagnostic tools intended for public health emergencies. BARDA's Broad Spectrum Antimicrobials Program had a 2017 budget of $192 million specifically for establishing public–private partnerships that develop novel antibiotic products (US Department of Health and Human Services, 2017). BARDA currently has at least seven different antibiotic R&D public–private partnerships with both large pharmaceutical companies, such as GSK, Pfizer and Roche, as well as several SMEs (Access to Medicine Foundation, 2018). BARDA is unique in that it offers ongoing push funding and guidance to all its partners, as well as the possibility for pull-based purchasing commitments for select marketable antibiotics. Jointly offered by the NIH and BARDA, the Antimicrobial Resistance Diagnostic Challenge offers a $20 million market entry reward to a developer of a rapid point-of-care diagnostic test that can aid in identifying antibiotic-resistant pathogens (US National Institutes of Health, 2016).

The Bill & Melinda Gates Foundation has been a key funder and partner of AMR initiatives that benefit low- and middle-income countries (LMICs), where the health burden of antibiotic resistance is greatest. The Bill & Melinda Gates Foundation has at least eight active R&D antimicrobial partnership projects for treating bacteria, tuberculosis, HIV, and malaria (Access to Medicine Foundation, 2018). Notably, in 2018, the Foundation committed $25 million to CARB-X (CARB-X, 2018a; 2018b).

UK initiatives

The majority of the UK-based antibiotic R&D initiatives were operated through the UK Research Councils, now brought together under the umbrella of UK Research and Innovation. These initiatives include the Cross-Research Council AMR Initiative, the Newton Fund, and the Global Challenge Research Fund. The Cross-Research Council AMR Initiative promotes a multidisciplinary approach to tackling AMR and offers a range of individual and collaborative grants to academic institutions. The initiative aims to break down research silos and involve LMICs in AMR research. To date, this initiative has committed approximately £50 million towards various AMR projects that target the earliest stages of the antibiotic value chain (UK Medical Research Council, 2016; Simpkin et al., 2017). The Newton Fund aims to strengthen scientific research partnerships between the UK and LMICs. The UK Research Councils alongside government agencies from China, India, and South Africa have pooled approximately £13.5 million in the Newton Fund for collaborative academic research on AMR (Newton Fund, n.d.; Simpkin et al., 2017). Finally, the recently established Global Challenge Research Fund is a £1.5 billion fund that will strive to address a multitude of challenges faced by LMICs. AMR is one of the key issues proposed for action through this fund (Research Councils UK, n.d.).

The UK National Institute for Health Research (NIHR) is another UK agency that offers support for antibiotic R&D. The NIHR's Biomedical

Research Centres and Health Protection Research Units have started a variety of programmes conducting basic science research that could lay the groundwork for antibiotic development (UK National Institute for Health Research, n.d.).

The Wellcome Trust has been an early champion for combating AMR and has financed numerous international antibiotic R&D initiatives. The Trust funded and hosted the Review on AMR, which led to the establishment of GAMRIF. Additionally, the Trust is a major funder of CARB-X and GARDP. As of 2018, the Wellcome Trust is a partner in at least 11 active public–private R&D projects for therapies targeting resistant bacteria, HIV and malaria (Access to Medicine Foundation, 2018).

Regulatory initiatives

Most antimicrobial agents are authorized in Europe through the centralized procedure of the European Medicines Agency (EMA) (Mossialos et al., 2010; Renwick, Simpkin & Mossialos, 2016). The EMA supports antibiotic developers through the licensing process by offering scientific advice and protocol assistance. Antibiotics can be assessed by the EMA via an expedited pathway to speed up possible market entry. Additionally, antibiotics that address unmet medical need may be granted conditional market authorization. These antibiotics are approved under weaker criteria for quality, safety, and efficacy to hasten patient access, but have much narrower indications for use and are reserved for those individuals without other treatment options. Some antibiotics against rare pathogens may also be eligible to receive orphan drug designation and an associated market exclusivity extension.

The US Food and Drug Administration (FDA) offers similar incentives to antibiotic developers through the Qualified Infectious Diseases Products (QIDP) designation and the Limited Population Antibacterial Drug (LPAD) designation (Renwick, Simpkin & Mossialos, 2016). Novel antibiotics that qualify for QIDP status receive regulatory guidance from the FDA, priority review, and fast track consideration when being assessed for market approval. Certain QIDP antibiotics may also be eligible for a market exclusivity extension of five years. Antibiotics that target rare and deadly pathogens could be eligible for LPAD designation, which permits a streamlined and conditional approval process so that patients lacking appropriate treatment can receive early access to a promising novel antibiotic. Analogous to the EMA's conditional market authorization process, antibiotics with LPAD designation are studied using smaller clinical populations and would only be approved for a narrow indication limited to the in-need patient cohort. The FDA also has an orphan drug licensing programme that offers market exclusivity extensions among some other benefits (Mossialos et al., 2010).

The Transatlantic Task Force on Antimicrobial Resistance (TATFAR) is an international partnership bringing together health policy and regulatory agencies from the EU, the USA, Norway and Canada. TATFAR's key goal is knowledge exchange and coordination across the various partner agencies (Renwick, Simpkin & Mossialos, 2016). Through TATFAR, the EMA and the FDA have been working collaboratively to improve and align the market authorization processes for antibiotics in Europe and the USA*. Since late 2016, the EMA and the FDA have been working with the Japanese Pharmaceuticals and Medical Devices Agency (PMDA) to encourage and accelerate development of novel antibiotics. These agencies have recently agreed to harmonize their data requirements for certain aspects of clinical trials for new antibiotics (European Medicines Agency, 2016).

Next steps in global antibiotic development incentivization

The extensive array of antibiotic R&D incentives is commendable, and strides have been made towards reviving the antibiotics pipeline. However, the current incentive package has major gaps and deficiencies that inhibit the transition from basic science research all the way to bedside access. The end goal should be a continuum of incentivization that reflects the economic need, cost distribution, and barriers of the entire antibiotic value chain. Different types of incentives are better suited for tackling different stages of this value chain (Figure 6.4). To achieve this continuum, there is a need to adjust push incentivization to increase funding of preclinical and clinical development, support global regulatory harmonization and provide added legal or regulatory incentives to facilitate market approval. There is also a need to introduce a variety of outcome-based pull incentives to ensure the commercialization and distribution of licensed antibiotics. These incentive changes must involve inter-initiative coordination and be made within the context of broader public health goals related to sustainability, patient access, and medical need.

---

* The Canadian and Norwegian drug regulatory agencies are not yet TATFAR partners.

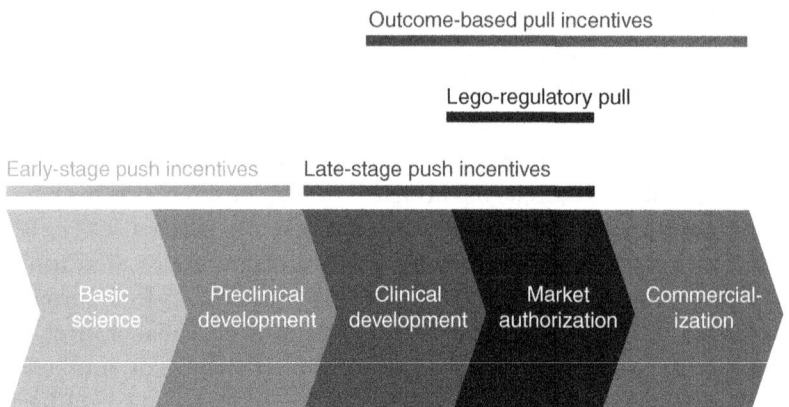

Figure 6.4 Continuum of incentivization across the antibiotic value chain
Source: Adapted from Simpkin et al., 2017

Push incentives, such as grants for researchers and direct project funding, are best used to facilitate the earlier stages of R&D from basic science up to clinical development. Most push funding for antibiotic R&D is directed towards basic antimicrobial science and less so towards clinical development. An estimated 86% of European national-level public funding of antibiotics was in this category (Kelly et al., 2015). The JPIAMR, DG-RTD, CARB-X, NIAID, UK Research and Innovation, and NIHR, preferentially fund antimicrobial basic science. While early-stage push funding of antimicrobial science is integral to the R&D process, there is a need for more late-stage push funding of preclinical and clinical trials to help translate scientific innovation into marketable products. The overemphasis on early-stage push funding reflects the fact that basic science lends itself more easily to being partitioned into projects requiring smaller individual monetary commitments than clinical trials do. In addition, public funders can more easily justify supporting nonprofit academic work. Basic science is largely the domain of academia and, as a result, private companies often do not benefit from early-stage push funding. Yet clinical trials, which are usually operated by private companies, are by far the most expensive aspect of R&D (O'Neill, 2015a). SMEs are the most impacted by the lack of late-stage push funding as they often struggle to raise the capital necessary for clinical trials (Renwick, Brogan & Mossialos, 2016).

As more drug candidates transition to clinical development, early-stage push funding could be pooled and reallocated to late-stage push funding to ensure viable antibiotics make it to the market approval stage. In addition, programmes such as BARDA and the IMI's COMBACTE, which specifically fund clinical trials, could be further expanded. As demonstrated in the WHO pipeline analysis, most antibiotics and alternative therapies are in phase I clinical trials and could immediately benefit from late-stage push funding (World Health Organization, 2017b). It will be important to balance this shorter-term strategy with the need to maintain a steady inflow of novel drug candidates identified through early discovery programmes.

Lego-regulatory (legal or regulatory) pull incentives, like those offered by the EMA and FDA, are most effective at facilitating progress through the market approval stage. Both the EMA and FDA offer several incentives that decrease the approval timeline for antibiotics: regulatory guidance, expedited pathways, and conditional market authorization. The primary value of these incentives comes from indirectly increasing the effective patent period of the antibiotic since it reaches the market earlier (Mossialos et al., 2010). But, there is a balance to be struck between speeding up the approval process and ensuring that licensed drugs meet standards for quality, safety, and efficacy (Renwick, Simpkin & Mossialos, 2016). It is unlikely that these regulatory processes can be shortened any further without sacrificing regulatory standards. In addition, many of the pipeline antibiotics are not expected to be high-volume products and therefore adding to their effective patent period does not translate into meaningful revenue. Market exclusivity extensions suffer from this same problem. Therefore, it may be worthwhile to explore alternative incentives that allow priority review (e.g. priority review vouchers (PRVs)) or market exclusivity extensions (e.g. transferable intellectual property rights (TIPRs)) to be transferred from an approved antibiotic to another product in the developer's portfolio that would benefit more from the longer effective patent period (Ferraro, Towse & Mestre-Ferrandiz, 2017). Incentives such as PRVs and TIPRs could provide a market incentive to license new antibiotics without requiring upfront government funding. However, it is important to be aware that PRVs and TIPRs do not incentivize antibiotic commercialization and they could have broader pharmaceutical market consequences (Mossialos et al., 2010; Ferraro, Towse & Mestre-Ferrandiz, 2017).

Harmonization between the EMA and the FDA's market approval requirements has been a step towards lowering market approval costs and time. However, the EMA and FDA regulatory processes are relatively similar unlike the Japanese PMDA or Chinese Food and Drug Administration. Harmonization efforts among these agencies will prove more challenging but could further relieve companies of duplicative regulatory approval costs. Including the PMDA in TATFAR was a laudable starting-point.

Push funding and legal or regulatory incentives can drive viable antibiotics to licensing; however, they are weak incentives for the commercialization and distribution of the product. Net profits from sales of an innovative new antibiotic are perceived to be limited for several reasons, especially when compared to therapeutic areas with the highest sales revenues; for example, oncologic, anti-diabetic, and anti-rheumatic drugs (EvaluatePharma, 2017). A novel antibiotic would be reserved as a last resort or may only target a rare resistant pathogen, which restricts potential sales revenue. High product prices are unlikely to compensate for low sales volume because of the considerable overlap in effectiveness between existing antibiotics. Also, future rapid-point-of-care diagnostic tools could cut into the revenue potential for newly marketed antibiotics (Outterson et al., 2015). Therefore, large outcome-based pull incentives are necessary in the absence of a viable market. Pull incentives have the added benefit of potentially allowing SMEs to secure venture capital for clinical trials. However, pull incentives for antibiotics have been mostly absent from current funding initiatives. The only outcome-based pull incentives currently available are relatively limited advanced market commitments (AMCs) offered by BARDA and GARDP for certain low-volume antibiotics (Simpkin et al., 2017).

Market entry rewards (MERs) have repeatedly been recommended by major reports and journal articles as an effective pull incentive for antibiotic commercialization (Ferraro, Towse & Mestre-Ferrandiz, 2017; Rex & Outterson, 2017; Simpkin et al., 2017; O'Neill, 2016; Renwick, Simpkin & Mossialos, 2016; Chorzelski et al., 2015). A MER is a financial prize for the successful development and licensure of an innovative antibiotic. To receive the prize, a developer must ensure that the antibiotic meets predefined product criteria and adheres to postmarket authorization conditions related to sustainability and patient access as specified by the payer. It is expected that a MER would need to be approximately $1–2 billion per first-entrant novel antibiotic to entice developers to invest in R&D and gamble on inventive antibiotic projects (DRIVE-AB, 2018). Practically, a prize of this size might be paid out as instalments over five to seven years. This would create a guaranteed revenue stream for the developer, spread out payer expenditures, and provide the payer with leverage if the developer chose to deviate from the agreed MER conditions. MERs can also be designed to have varying degrees of delinkage. Delinkage, in the context of a MER, refers to how much of a MER winner's revenue can be generated from sales volume (Rex & Outterson, 2016). A fully-delinked MER would pay for the antibiotic patent or licence in return for access to the drug at the cost of production. A partially delinked MER would still allow developers to generate some revenue from antibiotic sales. A fully delinked MER would thus need to be much larger than a partially delinked MER. Numerous other design variations, stipulations, and augmentations can be applied to the basic MER model to achieve various market goals. However, this is beyond the scope of this chapter. Both the 2018 DRIVE-AB final report and the 2017 Office of Health Economics report offer in-depth discussion of and recommendations for MER design and costing (DRIVE-AB, 2018; Ferraro, Towse & Mestre-Ferrandiz, 2017).

The key barrier to implementing a MER programme is the cost. With the 10-year goal of bringing 10 to 15 novel antibiotics to market, a MER programme is estimated to cost between $10 and $30 billion (DRIVE-AB, 2018; Ferraro et al., 2017; O'Neill, 2015a). Such a MER programme would provide large payouts of $1–2 billion for first entrants and increasingly smaller prizes for follow-up therapeutic products. A MER fund of this scale can only be practically achieved by pooling financial commitments from numerous countries and institutions into a ring-fenced endowment. For a MER programme to effectively pull antibiotics to the market, it is important that developers perceive this fund to be guaranteed by participating governments and protected from other public expenditures. This type of international fund for MERs has been recommended by various journal articles and international reports, such as the UK's AMR review, the Boston Consulting Group's report for the GUARD initiative, and DRIVE-AB (DRIVE-AB, 2018; Hoffman et al., 2015; Renwick, Brogan & Mossialos, 2016; Rex & Outterson, 2016; Stern et al., 2017; O'Neill, 2016). Despite the abundance of expert literature calling for an international MER programme, no nation has been willing to take the lead in establishing such a global fund or make a firm financial commitment. This inaction stems from the large sums involved, insufficient political support, the complexity of coordinated action, and a lack of capacity and expertise to implement such a scheme. In lieu of a global MER programme, alternative outcome-based pull incentives could be applied such as corporate tax incentives, value-based pricing and reimbursement strategies, and national AMCs for bulk purchasing (Renwick, Simpkin & Mossialos, 2016). These strategies are generally weaker incentives but do not require the same upfront financial commitment as a MER programme and thus may be more politically palatable.

Global cooperation and communication will be essential to creating the described continuum of antibiotic incentivization. Presently, national governments, global institutions, nongovernmental organizations, and industry are independently investing their resources in antibiotic R&D projects and funding programmes (Renwick, Simpkin & Mossialos, 2016). This is partially responsible for the current mismatched and incomplete global incentives. In addition, many of the antibiotic R&D initiatives operate in isolation from other initiatives despite their commonalities. There is a clear risk of duplicating efforts with initiatives that have similar mandates and receive interweaving funding from different payers. Therefore, there is a need for a global governing body that can coordinate antibiotic R&D incentive programmes at an international level and guide their operation at national levels. This global governing body could establish a unified direction for international antibiotic R&D incentives and guide incentive programmes towards achieving a more balanced global R&D incentive profile. Having such an entity would also help ensure that broader AMR goals related to global sustainability, patient access, and medical need are reinforced by the individual incentives.

Other recent reports such as the UK Review on Antimicrobial Resistance (O'Neill, 2016) have also referred to the need for a global body to coordinate, prioritize and mobilize resources for fighting AMR without defining how this might be established and what form it might take. The most concrete proposal emerged from the G20 summit in Hamburg in 2017 where G20 leaders called for "a new international R&D Collaboration Hub to maximize the impact of existing and new antimicrobial basic and clinical research initiatives as well as product development" (G20 Leaders' Declaration, 2017). The Global Antimicrobial Resistance Collaboration Hub is now being established in Germany with support from the Bill and Melinda Gates Foundation and the Wellcome Trust but requires political and financial support from many countries if it is to become an effective international instrument against AMR. It is intended that the Hub will coordinate efforts to promote antimicrobial research and encourage global involvement and investment and that its scope will include all stages of the antimicrobial development pipeline, as well as vaccines, alternative therapies and new diagnostic tools. It will be open to all countries and to nongovernment donors. Members will be expected to release additional investment in national and/or international research, but there will not be a set tariff for involvement.

Conclusion

Adding innovative antibiotics to the treatment arsenal is a critical aspect to addressing the AMR crisis. Incentives are necessary for overcoming the multitude of scientific, regulatory, and economic barriers that impede progress through the entire antibiotic value chain. Over recent years, many international, European, and national-level incentive programmes have been implemented to foster the antibiotic value chain. These have helped to lift the clinical pipeline for antibiotics out of dormancy. However, the recent progress in R&D is not nearly sufficient to counteract the rapid advancement of resistance rates. The current global incentive package could be improved by ensuring that a continuum of incentives is offered to developers, reflecting the economic need, cost distribution, and barriers of the antibiotic value chain. A global governing body that provides overarching guidance to international and national-level incentive programmes will be necessary to achieving such a continuum and the establishment of the Global Antimicrobial Resistance Collaboration Hub is a promising initiative to make such a governing body a reality.

References

Access to Medicine Foundation (2018). Antimicrobial resistance benchmark 2018. Amsterdam: Access to Medicine Foundation. (https://accesstomedicinefoundation.org/publications/2018-antimicrobial resistance-benchmark/, accessed 06 September 2018).

Butler MS, Blaskovich MA, Cooper MA (2013). Antibiotics in the clinical pipeline in 2013. J Antibiot. 66(10):571–590.

Chorzelski S, Grosch B, Rentmeister H et al. (2015). Presentation: Breaking through the wall: Enhancing research and development of antibiotics in science and industry. Boston: Boston Consulting Group. (http:// docplayer .net/8855152-Breaking-through-the-wall.html, accessed 06 September 2018).

Combating Antibiotic Resistant Bacteria (CARB-X) (2018a). Progress against superbugs: Annual report 2017–2018. Boston, MA: Combating Antibiotic Resistant Bacteria. (https://carb-x.org/wpcontent/uploads/2018/01/2016_ CARB-X-Annual_Report.pdf, accessed 06 September 2018).

Combating Antibiotic Resistant Bacteria (CARB-X) (2018b). UK Government and Bill & Melinda Gates Foundation join Carb-X partnership in fight against superbugs. Boston, MA: Combating Antibiotic Resistant Bacteria. (https://carb-x.org/carb-x-news/uk-government-and-billmelinda-gates-foundation-join-carb-x-partnership-in-fight-againstsuperbugs/, accessed 06 September 2018).

Cooper C, Khan Mirzaei M, Nilsson AS (2016). Adapting drug approval pathways for bacteriophage-based therapeutics. Front Microbiol. 7:1209.

Czaplewski L, Bax R, Clokie M et al. (2016). Alternatives to antibiotics – a pipeline portfolio review. Lancet Infect Dis. 16(2):239–251.

DRIVE-AB (2016). Incentives to stimulate antibiotic innovation: the preliminary findings of DRIVE-AB. DRIVE-AB. (http://drive-ab.eu/ wp-content/uploads/2016/06/WP2-Prereading-FINAL.pdf, accessed 06 September 2018).

DRIVE-AB (2018). Revitalizing the antibiotic pipeline: Stimulating innovation while driving sustainable use and global access. DRIVE-AB. (http://drive-ab.eu/wp-content/uploads/2018/01/DRIVE-AB-Final-Report-Jan2018.pdf, accessed 06 September 2018).

Eckhouse S, Sullivan R (2006). A survey of public funding of cancer research in the European Union. PLoS Med. 3(7):e267.

European and Developing Countries Clinical Trials Partnership (2017). EDCTP Annual report 2016. The Hague: European and Developing Countries Clinical Trials Partnership. (http://www.edctp.org/publication/ edctp-annual-report-2016/, accessed 06 September 2018).

European Commission (2015). European Commission launches €1m prize for a diagnostic test to combat antibiotic resistance. Brussels: European Commission. (http://ec.europa.eu/research/index .cfm?pg=newsalert&year=2015&na=na-260215, accessed 06 September 2018).

European Investment Bank (n.d.). What is InnovFin Infectious Diseases? Luxembourg: European Investment Bank. (http://www.eib.org/attachments/ thematic/innovfin_infectious_diseases_en.pdf, accessed 06 September 2018).

European Investment Bank (2017). EIB grants Da Volterra EUR 20m loan to speed up development of innovative solutions for the prevention and treatment of antibiotic-resistant infections. Luxembourg: European Investment Bank. (http://www.eib.org/infocentre/press/releases/all/2017/2017-249-la-bei-finance-da-volterra-pour-accelerer-ledeveloppement-de-ses-solutions-innovantes, accessed 06 September 2018).

European Medicines Agency (2016). Tripartite meeting between EMA, PMDA and FDA on regulatory approaches for the evaluation of antibacterial agents. London: European Medicines Agency. (http:// www.ema.europa .eu/docs/en_GB/document_library/Other/2016/09/ WC500212649.pdf, accessed 06 September 2018).

EvaluatePharma (2017). World Preview 2017, Outlook to 2022. London: EvaluatePharma. (http://info.evaluategroup.com/rs/607-YGS-364/ images/ WP17.pdf, accessed 06 September 2018).

Ferraro JS, Towse A, Mestre-Ferrandiz J (2017). Incentives for new drugs to tackle anti-microbial resistance. London: Office of Health Economics. (https://www.ohe.org/publications/incentives-new-drugs-tackle-anti-microbial-resistance#, accessed 06 September 2018).

G20 Leaders' Declaration (2017). Shaping an interconnected world. Hamburg: G20 Germany 2017. (https://www.g20.org/profiles/g20/modules/custom/ g20_beverly/img/timeline/Germany/G20-leadersdeclaration.pdf, accessed 06 September 2018).

Global Antibiotic Research and Development Partnership (2017). More than EUR 56 million raised to fight antibiotic resistance. Geneva: Global Antibiotic Research and Development Partnership. (https://gardp.org/ news-and-resources/updates/ accessed 05 December 2019).

Hoffman SJ, Caleo GM, Daulaire N et al. (2015). Strategies to achieving global collective action on antimicrobial resistance. Bulletin World Health Organization. 93:867–876.

Innovative Medicines Initiative (n.d.). New drugs for bad bugs (ND4BB). Innovative Medicines Initiative Ongoing. Brussels: Innovative Medicines Initiative. (http:// www.imi.europa.eu/content/nd4bb, accessed 06 September 2018).

Joint Programming Initiative on Antimicrobial Resistance (2018). Supported projects. Stockholm: Joint Programming Initiative on Antimicrobial Resistance. (http://www.jpiamr.eu/supportedprojects/, accessed 06 September 2018).

Kelly R, Zoubiane G, Walsh D et al. (2015). Public funding for research on antibacterial resistance in the JPIAMR countries, the European Commission, and related European Union agencies: A systematic observational analysis. Lancet Infect Dis. 16(4):431–440.

Kelly R, Ward R, Goossens H (2016). Public funding for research on antibacterial resistance in the JPIAMR countries, the European Commission, and related European Union agencies: a systematic analysis. Lancet Infect Dis. 16:431–440.

Marks & Clerk (2015). From rare to routine: Life sciences report 2015 on medicines for rare diseases, vaccines and antibiotics. London: Marks & Clerk. (https://www.marks-clerk.com/MarksClerk/media/MCMediaLib/PDF's/Reports/Life-Sciences-Report-2015-From-rareto-routine.pdf, accessed 06 September 2018).

Mossialos E, Morel CM, Edwards S et al. (2010). Policies and incentives for promoting innovation in antibiotic research. Copenhagen: WHO Regional Office for Europe on behalf of the European Observatory on Health Systems and Policies. (http://www.euro.who.int/__data/ assets/pdf_file/0011/120143/E94241.pdf, accessed 06 September 2018).

Newton Fund (n.d.). Newton Fund [website]. (http://www.newtonfund.ac.uk/, accessed 06 September 2018).

O'Neill J (2015a). Securing new drugs for future generations: the pipeline of antibiotics. The Review on Antimicrobial Resistance. London: Wellcome Trust and Government of the United Kingdom. (http://amr- review.org/sites/default/files/SECURING%20NEW%20DRUGS%20 FOR%20FUTURE%20 GENERATIONS%20FINAL%20WEB_0. pdf, accessed 06 September 2018).

O'Neill J (2015b). Tackling a global health crisis: initial steps. The Review on Antimicrobial Resistance. London:Wellcome Trust and Government of the United Kingdom. (https://amr-review.org/sites/default/files/ RARJ3003_Global_health_crisis_report_20.03.15_OUTLINED. pdf, accessed 06 September 2018).

O'Neill J (2016). Tackling drug-resistant infections globally: final report and recommendations. The Review on Antimicrobial Resistance. London: Wellcome Trust and Government of the United Kingdom. (https://amr-review.org/sites/default/files/160518_Final%20paper_with%20 cover.pdf, accessed 06 September 2018).

Outterson K, Powers JH, Daniel GW et al. (2015). Repairing the broken market for antibiotic innovation. Health Aff. 34(2):277–285.

Pew Charitable Trusts (2016). A scientific roadmap for antibiotic discovery. Philadelphia: Pew Charitable Trusts. (http://www.pewtrusts.org/~/media/assets/2016/05/ascientificroadmapforantibioticdiscovery.pdf, accessed 06 September 2018).

Pew Charitable Trusts (2017). Assessment of nontraditional products in development to combat bacterial infections. Philadelphia: Pew Charitable Trusts. (https://www.pewtrusts.org/en/research-and-analysis/issue-briefs/2017/12/assessment-of-nontraditional products-in-development-to-combat-bacterial-infections, accessed 06 September 2018).

Pew Charitable Trusts (2018). Tracking the global pipeline of antibiotics in development. Philadelphia: Pew Charitable Trusts. (https://www.pewtrusts.org/en/research-and-analysis/issue-briefs/2014/03/12/ tracking-the-pipeline-of-antibiotics-in-development, accessed 06 September 2018).

Renwick MJ, Brogan DM, Mossialos E (2016). A systematic review and critical assessment of incentive strategies for discovery and development of novel antibiotics. J Antibiot. 69:73–88.

Renwick MJ, Simpkin V, Mossialos E (2016). Targeting innovation in antibiotic drug discovery and development: The need for a One Health – One Europe – One World Framework. Copenhagen: WHO Regional Office for Europe. (http://www.euro.who.int–data/assets/pdf_file/0003/315309/Targeting-innovation-antibiotic-drug-dand-d-2016.pdf, accessed 06 September 2018).

Renwick M, Mossialos E (2018). What are the economic barriers of antibiotic R&D and how can we overcome them? Expert Opin Drug Discov. 13(10):889–892.

Research Councils UK (n.d.). Global Challenges Research Fund. Swindon: Research Councils UK. (http://www.rcuk.ac.uk/funding/gcrf/, accessed 06 September 2018).

Rex JH, Outterson K (2016). Antibiotic reimbursement in a model delinked from sales: a benchmark-based worldwide approach. Lancet Infect Dis. 16(4):500–505.

Silver LL (2011). Challenges of Antibacterial Discovery. Clin Microbiol Rev. 24(1):71–109.

Simpkin VL, Renwick MJ, Kelly R et al. (2017). Incentivizing innovation in antibiotic discovery and development: progress, challenges and next steps. J Antibiot. 70(12):1087–1096.

So AD, Gupta N, Brahmachari SK et al. (2011). Towards new business models for R&D for novel antibiotics. Drug Resist Updat. 14(2):88–94.

So AD, Ruiz-Esparza Q, Gupta et al. (2012). 3Rs for innovating novel antibiotics: sharing resources, risks, and rewards. BMJ. 344:e1782.

Stern S, Chorzelski S, Franken L et al. (2017). Breaking through the wall: a call for concerted action on antibiotics research and development. Berlin: German Federal Ministry of Health. (https://www.bundesgesundheitsministerium.de/

fileadmin/Dateien/5_ Publikationen/Gesundheit/Berichte/GUARD_Follow_ Up_Report_ Full_Report_final.pdf, accessed 06 September 2018).

Thomas D, Wessel C (2015). BIO's Venture funding of therapeutic innovation. Washington, DC: BIO. (https://www.bio.org/biovcstudy, accessed 06 September 2018).

UK Government (2016a). Collaboration on health boosted at UK–China high level people to people dialogue. London: UK Government. (https:// www.gov.uk/government/news/collaboration-on-health-boosted-at-uk-china-high-level-people-to-people-dialogue. accessed 06 September 2018).

UK Government (2016b). Expert advisory board to support the Global AMR Innovation Fund. London: UK Government. (https://www.gov .uk/government/news/expert-advisory-board-to-support-the-global-amr-innovation-fund, accessed 06 September 2018).

UK Government (2018). £30 million of funding to tackle antimicrobial resistance. London: UK Government. (https://www.gov.uk/government/ news/30-million-of-funding-to-tackle-antimicrobialresistance, accessed 06 September 2018).

UK Medical Research Council (2016). Tackling AMR – A cross council initiative. London: Medical Research Council. (http://www.mrc.ac.uk/ research/initiatives/antimicrobial-resistance/tackling-amr-across-council-initiative/, accessed 06 September 2018).

UK National Institute for Health Research (n.d.). Research and impact: antimicrobial resistance. London: National Institute for Health Research. (https://www.nihr.ac.uk/research-and-impact/research-priorities/ antimicrobial-resistance.htm, accessed 06 September 2018).

US Centers for Disease Control and Prevention (2013). Antibiotic resistant threats in the United States, 2013. (https://www.cdc.gov/drugresistance/ threat-report-2013/pdf/ar-threats-2013-508.pdf, accessed 06 September 2018).

US Department of Health and Human Services (2017). HHS FY 2018 budget in brief – PHSSEF. Washington, DC: Department of Health and Human Services. (https://www.hhs.gov/about/budget/fy2018/ budget-in-brief/phssef/ index.html, accessed 06 September 2018).

US National Institutes of Health (2016). Antimicrobial Resistance Diagnostic Challenge. Washington, DC: National Institutes of Health. (https:// dpcpsi .nih.gov/AMRChallenge, accessed 06 September 2018).

US National Institutes of Health (2017). Estimates of funding for various research, condition, and disease categories (RCDC). Washington, DC:

National Institutes of Health. (https://report.nih.gov/categorical_spending .aspx, accessed 06 September 2018).

World Health Organization (2017a). Global priority list of antibiotic-resistant bacteria to guide research, discovery, and development of new antibiotics. Geneva: World Health Organization. (http://www.who.int/medicines/ publications/WHO-PPL-Short_Summary_25Feb-ET_NM_WHO.pdf, accessed 06 September 2018).

World Health Organization (2017b). Antibacterial agents in clinical development. Geneva: World Health Organization. (http://www.who .int/ medicines/news/2017/IAU_AntibacterialAgentsClinicalDevelopment_ webfinal_2017_09_19.pdf?ua=1, accessed 06 September 2018).

7 | Ensuring innovation for diagnostics for bacterial infection to combat antimicrobial resistance

ROSANNA W. PEELING, DEBRAH BOERAS, JOHN NKENGASONG

Introduction

At the Sixty-Eighth World Health Assembly in May 2015, Member States of the World Health Organization (WHO) endorsed a Global Action Plan to tackle antimicrobial resistance (AMR), the most urgent of which is antibiotic resistance (World Health Organization, 2015a). The goal of the Global Action Plan is to ensure continuity of successful treatment and prevention of infectious diseases with effective and safe medicines that are quality-assured, used responsibly, and accessible to all who need them. To achieve this goal, the Global Action Plan sets out five strategic objectives:

- to improve awareness and understanding of antimicrobial resistance;
- to strengthen knowledge through surveillance and research;
- to reduce the incidence of infection;
- to optimize the use of antimicrobial agents; and
- to develop the economic case for sustainable investment that takes account of the needs of all countries, and increase investment in new medicines, diagnostic tools, vaccines and other interventions.

Development of this plan was guided by the advice of countries and key stakeholders, based on several multi-stakeholder consultations at different global and regional forums. Diagnostics underpin all but the first of these strategic objectives.

The plan now requires rapid innovation, political will and buy-in from communities to succeed. This chapter will provide an overview of the unique challenges that the developers of diagnostics devices face, discuss policy options and tools that may help overcome such barriers, and discuss how economic tools such as economic assessment, may produce evidence to support policy-making.

155

Innovation in diagnostics to combat AMR

A global AMR response will require diagnostics that are affordable and accessible, can be used at the point-of-care (POC), and can rapidly determine antimicrobial susceptibility. These tests are urgently needed to reduce inappropriate use of antibiotics, guide patient management for improved outcomes and provide much needed AMR surveillance.

Diagnostics for more targeted use of antibiotics

Studies and systematic reviews have shown that the majority of antibiotics are used in primary health care or sold over the counter in pharmacies. A study conducted in 48 primary health care settings in China showed that 53% of outpatients were prescribed antibiotics, of which only 39% were prescribed properly, while 78% of inpatients were prescribed antibiotics, of which only 25% were prescribed properly (Wang et al., 2014). In all, 55% of prescriptions were for two or more antibiotics. Antibiotics were most commonly prescribed for colds and acute bronchitis.

For tertiary care settings, a point prevalence survey of antimicrobial utilization in a Canadian teaching hospital conducted in 2012 showed that one or more antimicrobial agents were prescribed in 31% and 4% of acute care and long-term care patients, respectively (Lee et al., 2015). The most common indications were respiratory and urinary tract infections for both acute and long-term care patients.

Many of the antibiotics prescribed empirically in primary health care settings are for common infectious disease syndromes:

- fever
- flu-like illness
- pneumonia
- sexually transmitted infections
- enteric infections
- urinary tract infections.

For any diagnostic test to be effective in primary health settings, it needs to be simple to perform, rapid, affordable and accurate. This means providing a result in less than 15–20 minutes to be able to guide more targeted use of antibiotics (Okeke et al., 2011). Traditional diagnostic tests are designed to identify pathogens in specimens taken from the patient. However, the syndromes listed above can be caused by many bacterial, viral or in some cases, fungal pathogens. It would be difficult to develop a test that can identify the cause or causes of these syndromes. As a compromise, a simple rapid test that can be used to distinguish between bacterial and viral infections would potentially be useful to inform health care providers whether a prescription for antibiotics is warranted. Researchers have turned to the host markers that may be used for this purpose.

Syndrome-based POC diagnostics using host biomarkers

A systematic review of host markers that could be used to distinguish between bacterial and viral infections showed that over 112 host bio-markers have been evaluated and published between 2010 and 2015 (Kapasi et al., 2016). There was much heterogeneity between studies, including study outcomes, comparisons, spectrum of infections included in each group, methods for clinical and microbiological assessment, diseases/conditions and biomarkers tested, type of samples used, sites of infection and the quality of the studies. The study quality scores ranged from 23% to 92%, depending on the number of patients per strata, number of comparisons made and statistical correction, and blinding. Most studies were performed in high-income countries, with only 19% conducted in the developing world. The most frequently evaluated host biomarkers were C-reactive protein (CRP) (61%), white blood cell count (44%) and procalcitonin (34%). There were nine high performance host biomarkers or combinations, with sensitivity and specificity of >85% or 100% for either sensitivity or specificity (Table 7.1). Five host biomarkers were considered weak markers as they lacked statistically significant performance in discriminating between bacterial and nonbacterial infections.

Some of the high performing biomarkers have been commercialized as single or combination assays. These include ImmunoXpert™ (CRP+IP-10+TRAIL, CE-marked); FebriDx™ (MxA+CRP); and SeptiCyte (nondisclosed). Others are in the pipeline. None of these assays have yet achieved all the minimal or desired characteristics set out in the Target Product Profile (TPP) published by the Foundation for Innovative New Diagnostics (Dittrich et al., 2016).

Table 7.1 *High performing biomarkers for distinguishing between bacterial and viral infections*

BioMarkers	Type of biomarker (specimen)	Performance Sensitivity	Specificity	Number of studies – reference (quality score[a])
Respiratory infections:				
Procalcitonin + 10-gene classifier	Inflammatory + genetic (blood, adult)	95%	92%	1 – Suarez et al., 2015 (54%)
48-gene classifier	Genetic (blood, adult)	89%	94%	1 – Zaas et al., 2013 (85%)
IL-4	Cytokine (blood, adult)	100%	77%	2 – Haran et al., 2013; Burdette et al., 2014 (23–58%)
Meningitis:				
Heparin binding protein	Homeostasis (CSF)	100%	99%	2 – Linder et al., 2011; Chalupa et al., 2011 (42–62%)
Lactate	Metabolic (CSF, adult and paediatric)	94–96%	94–97%	3 – Linder et al., 2011; Viallon et al., 2011; Huy et al., 2011 (54–62%)
PMN counts	Haematological (CSF, adult)	93–96%	85–96%	4 – Linder et al., 2011; Ibrahim, Abdel-Wahab & Ibrahim, 2011; Abdelmoeaz et al., 2014; Chalupa et al., 2011 (46–65%)

Table 7.1 *(cont.)*

Bacterial versus viral infections:				
CRP+IP10+ TRAIL	Combination (blood, adult and paediatric)	95%	91%	1 – Oved et al., 2015 (92%)
CD35+cd32+ CD88+MHC-1	Cytological (blood, adult)	91%	92%	1 – Nuutila et al., 2013 (62%)
MxA	(Blood, paediatric)	87%	91%	1 – Kawamura et al., 2012 (39%)

Notes: CSF: cerebrospinal fluid; PMN: polymorphonuclear neutrophil; CRP: C-reactive protein; IP10: interferon-γ-induced protein; TRAIL: tumour necrosis factor-related apoptosis-inducing ligand; MxA: myxoma resistance protein 1.

[a]Studies were scored using 26 parameters from QUADAS; a score of >60% was considered high quality.

Source: Kapasi et al., 2016.

To stimulate interest in innovation in a simple, affordable, rapid diagnostic test that can be used at POC, several developed countries have set up challenge prizes. The first was the Horizon 2020 prize for better use of antibiotics for respiratory infections. The prize was awarded in February 2017 to the development of a neutrophil marker, human neutrophil lipocalin on the Philips Minicare platform (Horizon 2020, n.d). The test uses a single drop of blood from a finger-prick and takes less than 10 minutes to provide a result. The usefulness of this biomarker remains to be proven in large-scale clinical trials.

In 2015, the United Kingdom announced the Longitude Prize of £10 million. The challenge is to invent an affordable, accurate, fast and easy-to-use test for bacterial infections that will allow health professionals worldwide to administer the right antibiotics at the right time. The challenge is currently ongoing with the final submission due in September 2022 (Longitude Prize, n.d).

In September 2016, the US Department of Health and Human Services announced a challenge prize competition in which up to $20 million will be awarded for one or more novel and innovative POC diagnostics that would have clinical and public health value in combating the development and spread of antibiotic-resistant bacteria (National Institutes of Health, 2017).

POC diagnostics for pathogen detection and susceptibility testing

In 2013, the US Centers for Disease Control and Prevention (CDC) published a list of pathogens for which resistance poses different levels of threats to public health in the USA (Table 7.2). In 2017, the WHO published a list of bacteria for which drug research and development (R&D) is urgently needed that has many common elements with the CDC list. POC diagnostics developed for these infections may slow the spread of resistance.

Gonococcal resistance is considered an urgent threat on both lists. In a background paper for the UK's Review on Antimicrobial Resistance (O'Neill, 2016), a modelling study showed that the major benefit of POC testing for gonorrhoea is increasing the proportion of patients treated appropriately on the same day as testing (Turner et al., 2018). As POC tests with sufficient accuracy will normally cost more than laboratory-based high-throughput tests, policy-makers need to balance the additional cost with increased patient and system-level benefits. POC

Table 7.2 *Resistant pathogens posing public health threats as prioritized by the US Centers for Disease Control and Prevention*

Level of threat	Pathogens
Urgent	
	Clostridium difficile
	Carbapenem-resistant Enterobacteriaceae (CRE)
	Drug-resistant *Neisseria gonorrhoeae*
Serious	
	Multidrug-resistant *Acinetobacter*
	Drug-resistant *Campylobacter*
	Fluconazole-resistant *Candida*
	Extended spectrum beta-lactamase-producing Enterobacteriaceae (ESBLs)
	Vancomycin-resistant *Enterococcus* (VRE)
	Multidrug-resistant *Pseudomonas aeruginosa*
	Drug-resistant non-typhoidal Salmonella
	Drug-resistant *Salmonella typhi*
	Drug-resistant *Shigella*
	Methicillin-resistant *Staphylococcus aureus* (MRSA)
	Drug-resistant *Streptococcus pneumoniae*
	Drug-resistant tuberculosis

Table 7.2 *(cont.)*

Level of threat	Pathogens
Concerning	
	Vancomycin-resistant *Staphylococcus aureus* (VRSA)
	Erythromycin-resistant Group A *Streptococcus*
	Clindamycin-resistant Group B *Streptococcus*

Source: US Centers for Disease Control and Prevention, 2013.

Table 7.3 *WHO list of priority pathogens for R&D of antibiotics*

Priority	Resistance	Pathogens
Critical:		
	Carbapenem	*Acinetobacter baumannii*
		Pseudomonas aeruginosa
	+ cephalosporin	Enterobacteriaceae
High:		
	Vancomycin	*Enterococcus faecium*
	+ methicillin	*Staphylococcus aureus*
	Clarithromycin	*Helicobacter pylori*
	Fluoroquinolone	*Campylobacter*
		Salmonella spp.
	+ cephalosporin	*Neisseria gonorrhoeae*
Medium:		
	Penicillin	*Streptococcus pneumoniae*
	Ampicillin	*Haemophilus influenzae*
	Fluoroquinolone	*Shigella* spp.

Source: World Health Organization, 2017b.

tests have already been shown to improve patient outcomes as well as increasing the efficiency of the heallth care system by reducing number of patient visits (Mabey et al., 2012; García et al., 2013; Jani & Peter, 2013). In the case of a POC test for gonorrhoea, policy-makers must balance the cost of the POC test against improved patient outcomes and public health benefits. A simple POC test can potentially reduce the risk of onward transmission to a sexual partner, reduce loss to follow-up and potentially improve partner notification, and further reduce the reservoir of infection in the community.

A more critical innovation is to develop a test that would allow providers to discriminate between sensitive and resistant pathogens at POC, which would facilitate the re-introduction of abandoned first-line therapies. The modelling study for the AMR review estimated that if ciprofloxacin could be used in place of ceftriaxone in the 63% of individuals with ciprofloxacin-susceptible infections, this could save over 22 000 doses of ceftriaxone annually in the UK alone (Turner al., 2018). Reducing the use of antibiotics, especially of last-line therapies, is a key aim of the UK national strategy on antimicrobial resistance; being able to reuse older, cheaper drugs is an important economic benefit.

While it is encouraging that governments are stimulating technological innovations through induction or challenge prizes, and TPPs have been developed to guide test development, test developers still face many barriers in moving forward with these promising technologies (Peeling & Nwaka, 2011).

Barriers to innovation in diagnostics

Identifying the testing needs to respond to AMR is the first step in bringing urgently needed innovative diagnostics from the bench to the bedside. This pathway can be roughly divided into three phases, each driven by different players:

1) The R&D phase is driven by industry and test developers in the public sector, such as academia and research institutions. This phase can take anywhere from five to 10 years, with an investment ranging from $10 to 200 million.
2) There are two major players in the second phase – the test developers and the regulators. The test developers can spend two to three years conducting clinical trials in intended markets to gather data on how well the test performs for submission to the regulatory authorities. The regulatory review and audits of manufacturing quality can then take more than two years.
3) After a test receives regulatory approval, the third and final phase involves policy-makers, disease control programme managers and chiefs of laboratory services. These players should conduct a health technology assessment (HTA) of the new test to determine the potential clinical benefit and cost–effectiveness for their programme. If the results are favourable, policies are developed to define how the new test will be used, who will be allowed to perform it, who

will act on the results, and whether it will be reimbursable through public funds. This phase will also require authorized procurement and implementation. This can still take another three to five years.

Taken together, even if a promising diagnostic test is available for clinical trials today, it could take seven to 10 years, and millions of dollars, before it is widely adopted and used. Since diagnostics have a much shorter life-cycle than drugs or vaccines, the lengthy and fragmented pathway to market entry limits return on investment.

For diagnostic products with a viable commercial market, this pathway is driven, funded and managed largely by the private sector drawing on appropriate expertise as needed. For diagnostics of public health importance in the developing world, there is often little interest in investing in research for a pipeline of products that would be appropriate and useful due to a perceived lack of return on investment. Developers often have limited knowledge of the TPP and have difficulties obtaining specimens and reagents that can help them with test development and calibration. They have difficulties networking and negotiating with sites in developing countries for field trials, which delays the time to market (Yager et al., 2008; Chin, Linder & Sia, 2012; Kumar et al., 2015). The demand by many regulators for clinical trials in their own country has led to duplications of studies to evaluate test performance and utility, which further delays regulatory approvals and adds costs. Many countries in the developing world do not have the regulatory and HTA expertise to support policy development that will expedite regulatory approval and test adoption. Hence, the result is delayed and costly diagnostics and lack of overall systems to sustain these diagnostics (McNerney & Peeling, 2015; Rugera et al., 2014).

In recent years, efforts have been made to confront these barriers, particularly to combat AMR, by bringing together experts from different fields such as microbiologists, clinicians, engineers, regulators and policy-makers to share experiences and interact (Niemeier, Gombachika & Richards-Kortum, 2014; García et al., 2015; Derda et al., 2015). However, without leadership and sustained effort, this pathway remains fragmented with many gaps and challenges along the way. In summary, for an effective AMR response, innovation across several fronts is urgently needed to bring about a paradigm shift to accelerate and streamline the diagnostic pathway if promising POC tests are to be widely used to guide appropriate use of antibiotics in the foreseeable future (Box 7.1).

Diagnostics to conduct AMR surveillance

AMR surveillance is the cornerstone for assessing the burden of AMR and for providing data for action in support of local, national and global AMR strategies. Surveillance baseline data are critical for assessing the impact of interventions, such as stewardship. Surveillance also allows identification of emerging variants to inform further test development. AMR surveillance strategies can only be effective if the appropriate diagnostic tests are used for surveillance and the quality of the testing is assured. Surveillance data must also reach a decision-maker in a timely manner – and must be understandable, actionable, and then communicated to those who need to know.

One of the five strategic objectives of the WHO Global Action Plan is to strengthen the evidence base for AMR through enhanced global surveillance and research (World Health Organization, 2015a).

Box 7.1 Summary of diagnostic innovations urgently needed to reduce misuse of antibiotics

Technological innovation needs:

Simple and rapid biomarker or pathogen-based tests that can be used at the point-of-care to differentiate between bacterial and non-bacterial infections. In particular, tests are required that are fit for use in primary health care by a health care worker for patients presenting with common syndromes such as fever, respiratory infections and UTIs. It has been proposed that these POC tests need to have a diagnostic accuracy of 90–95% sensitivity and 80–90% specificity at a cost of less than $5 (Dittrich et al., 2016).

Facilitating technological innovation requires:

- Sustained sources of funding.
- Clear definition-of-use case scenarios and consensus on TPPs to guide test development.
- Equitable access to biobanks of well-characterized specimens to make it more attractive for developers to enter the development pathway.

Box 7.1 (cont.)

Innovations in policy development require:

- Regional regulatory harmonization on safety and effectiveness of the new tests to avoid duplication and accelerate approval and adoption across multiple countries.
- HTA capacity for countries in resource-limited settings. Involving the regulators in the HTA process so that the assessment of risk and benefit can be carried out simultaneously, instead of sequentially, to accelerate test adoption.
- Diagnostic algorithms on how to use the tests within a clinical pathway.

Innovation in delivery and financing needs:

- Efficient systems for training, supply chain management, quality assurance and monitoring safety and effectiveness.
- Financing mechanisms applicable to developing countries, similar to that of Gavi, the Vaccine Alliance.

Table 7.4 *Pathogen–antimicrobial combinations on which GLASS will collect data*

	Antibacterial class	Antibacterial agents that may be used for antimicrobial susceptibility testing
Escherichia coli	Sulfonamides & trimethoprim	Co-trimoxazole
	Fluoroquinolones	Ciprofloxacin or levofloxacin
	3rd generation cephalosporins	Ceftriaxone or cefotaxime and ceftazidime
	4th generation cephalosporins	Cefepime
	Carbapenems	Imipenem, meropenem, ertapenem or doripenem
	Polymyxins	Colistin
	Penicillins	Ampicillin

Table 7.4 *(cont.)*

	Antibacterial class	Antibacterial agents that may be used for antimicrobial susceptibility testing
Klebsiella pneumoniae	Sulfonamides & trimethoprim	Co-trimoxazole
	Fluoroquinolones	Ciprofloxacin or levofloxacin
	3rd generation cephalosporins	Ceftriaxone or cefotaxime and ceftazidime
	4th generation cephalosporins	Cefepime
	Carbapenems	Imipenem, meropenem, ertapenem or doripenem
	Polymyxins	Colistin
Acinetobacter baumannii	Tetracyclines	Tigecycline or minocycline
	Aminoglycosides	Gentamycin and amikacin
	Carbapenems	Imipenem, meropenem or doripenem
	Polymyxins	Colistin
Staphylococcus aureus	Penicillinase-stable beta-lactams	Cefoxitin
Streptococcus pneumoniae	Penicillins	Oxacillin, Penicillin G
	Sulfonamides & trimethoprim	Co-trimoxazole
	3rd generation cephalosporins	Ceftriaxone, cefotaxime
Salmonella spp.	Fluoroquinolones	Ciprofloxacin or levofloxacin
	3rd generation cephalosporins	Ceftriaxone or cefotaxime and ceftazidime
	Carbapenems	Imipenem, meropenem, ertapenem or doripenem
Shigella spp.	Fluoroquinolones	Ciprofloxacin or levofloxacin
	3rd generation cephalosporins	Ceftriaxone or cefotaxime and ceftazidime
	Macrolides	Azithromycin
Neisseria gonorrhoeae	3rd generation cephalosporins	Cefixime, ceftriaxone
	Macrolides	Azithromycin
	Aminocyclitols	Spectinomycin
	Fluoroquinolones	Ciprofloxacin
	Aminoglycosides	Gentamycin

Source: World Health Organization, 2015b.

The Global Antimicrobial Resistance Surveillance System (GLASS) has been launched to support a standardized approach to the collection, analysis and sharing of AMR data at a global level. These data can be used for decision-making and provide evidence for action and advocacy (Table 7.4).

GLASS aims to combine clinical, laboratory and epidemiological data on pathogens that pose the greatest threats to global public health. It is recognized that national surveillance systems will vary in levels of development and scale. Flexibility has therefore been built into the system to allow each country to participate from the outset while implementing and strengthening the core components of a national AMR surveillance system with a phased approach.

There are limited data on AMR surveillance in developing countries largely due to lack of access to diagnostics. Innovation in more affordable and user-friendly tests for surveillance at different levels of the health care system is urgently needed. Without a baseline of the extent of resistance, countries will not be able to measure the impact of their interventions, such as stewardship.

At the most basic level, countries can start conducting point prevalence surveys in hospitals. Point prevalence is the number of persons with disease in a time interval (e.g. one year) divided by the number of persons in the population; that is, prevalence at the beginning of an interval plus any incident cases. A point prevalence survey of antimicrobial use can be conducted on a specific day across an entire facility to provide baseline information on antibiotic usage and set potential targets for antibiotic stewardship (Lee et al., 2015).

For surveillance, the majority of commercially available molecular tests focus on detecting *Clostridium difficile*, methicillin-resistant *Staphylococcus aureus* (MRSA) and vancomycin-resistance markers. A few platforms also offer tests for carbapenem-resistant Enterobacteriaceae (CRE) and *Mycobacterium tuberculosis*. For commercially available MRSA assays, although they all show excellent sensitivity and specificity, they are all molecular assays that require two hours to complete and are too costly for most of the developing world. In general, the time to get results from molecular testing platforms ranges from less than one hour to five to eight hours. Details of AMR technologies can be found in three technology landscapes that have been published (University of Oxford, 2015; Global Antibiotic Research & Development Partnership, n.d.; UNITAID, 2018).

Box 7.2 Selected examples of AMR surveillance networks

Gonorrhoea resistance networks: Gonorrhoea is a sexually transmitted infection caused by *Neisseria gonorrhoeae*. In 2012, WHO estimated that there were 78 million cases worldwide. Since the introduction of antimicrobial treatment, resistance has rapidly emerged to Sulfonamides, penicillins, tetracyclines, macrolides, fluoroquinolones, and early-generation cephalosporins. Decreased susceptibility to ceftriaxone, the last-line treatment for gonorrhoea, has been reported from many, particularly well-resourced, settings globally. Dual therapy, mainly ceftriaxone plus azithromycin, is recommended. The WHO Global Gonococcal Antimicrobial Surveillance Programme is key to monitoring AMR trends, identifying emerging AMR, and informing refinements of treatment guidelines and public health policy globally. More information is available at: http://www.who.int/reproductivehealth/topics/rtis/gonococcal_resistance/en/.

Enter-net is an EU-wide network for the surveillance of human Salmonella and Verocytotoxin-producing *Escherichia coli* (VTEC) infections. By involving national reference laboratories and the epidemiologist responsible for national surveillance of these organisms, data from 15 countries are being collated every month to create international Salmonella and VTEC databases. More information is available at: http://ec.europa.eu/health/ph_projects/2000/com_diseases/ fp_commdis_2000_inter_01_en.pdf.

European Antimicrobial Resistance Surveillance Network (EARS-Net) is the largest publicly funded system for AMR surveillance in Europe. The objectives of EARS-Net are to: 1) collect comparable, representative and accurate AMR data; 2) analyse temporal and spatial trends of AMR in Europe; 3) provide timely AMR data for policy decisions; 4) encourage the implementation, maintenance and improvement of national AMR surveillance programmes; and 5) support national systems in their efforts to improve diagnostic accuracy by offering annual external quality assessments. More information is available at: https://ecdc.europa.eu/en/about-us/networks/disease-networks-and-laboratory-networks/ears-net-about.

Box 7.2 (cont.)

CDC's Foodborne Diseases Active Surveillance Network (FoodNet) includes 10 US sites and monitors cases reported caused by nine enteric pathogens commonly transmitted through food. FoodNet conducts active, population-based surveillance for laboratory-diagnosed infections caused by *Campylobacter*, *Cryptosporidium*, *Cyclospora*, *Listeria*, *Salmonella*, Shiga toxin-producing *Escherichia coli* (STEC), *Shigella*, *Vibrio* and *Yersinia*. In 2015, surveillance from these 10 sites covered an estimated 49 million people, representing 15% of the US population. Infections are confirmed by culture or culture-independent diagnostic tests detecting bacterial pathogen antigen, nucleic acid sequences, or for STEC, Shiga toxin or Shiga toxin genes, in a stool specimen or enrichment broth. More information is available at: https://www.cdc.gov/mmwr/volumes/66/wr/ mm6615a1.htm.

Respiratory Infection Networks: The Global Point Prevalence Survey of Antimicrobial Consumption and Resistance (GLOBAL-PPS) is an example of a respiratory infections network with participation from 73 countries. The network tracks the causes of respiratory infections and associated antibiotic consumption. Global-PPS also supports a point prevalence surveys (PPS) e-learning module to learn how to use to measure antibiotic consumption and fight antimicrobial resistance. More information is available at: http:// www.global-pps.com.

Africa CDC AMR Surveillance Network: In October 2017, the Africa Centres for Disease Control and Prevention (Africa CDC) launched its AMR surveillance network (AMRSNET). As part of the African Union, Africa CDC supports African countries to improve surveillance, emergency response, and prevention of infectious diseases. This includes outbreaks, man-made and natural disasters, and public health events of regional and international concern. It also seeks to build the capacity to reduce the disease burden on the continent. Africa CDC will work with African countries to develop policy frameworks for AMR surveillance. More information is available at: https://au.int/en/pressreleases/20171107/african-countries-launch-framework-tackle-threat-antibiotic-resistant.

Box 7.3 Summary of AMR surveillance innovation needed

- Robust and high-throughput assays for immediate pathogen identification to provide regional and country disease risk assessments and support global health decisions.
- Tests with data connectivity and GPS capability to promote timely information provision in support of resource allocation.
- Technological innovation to develop more affordable and user-friendly tests for surveillance in resource-limited settings.

Key issues to consider in biosurveillance include pathogen identification, sequence sharing, common clinical case definition, standardized assays and kit types, including standard operating procedures. A very important element of surveillance systems is the use of diagnostic devices that have location services (GPS), time/date stamps, and data transmission capabilities. Automated results and information sharing can prove useful to biosurveillance programmes. A survey of viral gastroenteritis outbreaks in Europe showed the difficulties of interpreting surveillance data when different diagnostic tests were used for reporting (Lopman et al., 2003).

The backbone of global biosurveillance will include AMR surveillance networks. A number of networks have been established and valuable lessons can be learnt from them.

POC diagnostics to decrease the cost of drug trials

Drug development is a lengthy and costly process with a huge "valley of death" along the developmental pathway. In recent years, drug companies have turned away from developing anti-infectives to developing drugs for chronic diseases, which offers a more consistent market and a longer time for return on investment. To incentivize drug companies to return to developing antibiotics, the public and private sectors should work in partnership to improve the efficiency and effectiveness of the process of bringing a drug to market. One of the major costs of bringing a drug to market is the cost of the clinical trials. It has been estimated that the use of a POC test to identify the target patient population early in a clinical trial can reduce time for enrolment and result in significant cost savings (Savuto & Karuppan, 2017).

The WHO has published a list of pathogens for which antibiotic R&D needs to be prioritized (Table 7.3). The development of POC diagnostics that can be used to accurately identify patients with these infections for drug study recruitment will significantly improve the efficiency of drug trials and help decrease the cost of trials compared to recruiting patients based on disease syndromes.

Antibiotics, used appropriately, will continue to play a critical part in modern medicine and public health. There are numerous opportunities to ensure appropriate antibiotic use through innovations for diagnostics. A faster regulatory approval process for diagnostics and a national policy framework for AMR will help countries combat AMR through testing and surveillance. Countries will still need to explore new sources of funding for procurement of tests and implementation of AMR diagnostics and programmes.

Lowering the cost of diagnostic R&D

A robust pipeline of diagnostics for AMR is needed to address the many different needs. Additional mechanisms to incentivize diagnostics R&D are required. Funding agencies can offer to de-risk investments for diagnostic R&D by offering loans that only need to be paid back if the company makes a profit on the product. Other possible mechanisms are to attract impact investments, leverage investments made to develop open platform technologies for epidemic preparedness, and to partner with vaccine and drug companies for R&D.

Lowering the cost of market entry and reducing delay

The regulatory approval processes for diagnostics are often lengthy, costly and not transparent. Regulation of medical products is intended to ensure safety and quality while balancing the need for timely access to beneficial new products. Current regulatory oversight of diagnostic tests in developing countries is highly variable (Rugera et al., 2014). While weak regulation allows poor-quality tests to enter the market, inefficient or overzealous regulation results in unnecessary delays, increases costs and acts as a barrier to innovation and market entry. Regulatory science lags far behind technological innovation (Morel et al., 2016). As a result, regulators are increasingly unable to assess the risk and benefit of novel technologies or are becoming increasingly risk-averse. Bringing together regulators, policy-makers, programme managers and subject matter experts as part of a HTA framework to assess jointly the risks and benefits of new technologies could ensure a fair and transparent assessment of risks and benefit and accelerate both regulatory approval and policy development.

A second solution for lowering the regulatory barrier to innovation is to set international standards for diagnostic evaluations similar to those developed for drugs and vaccines. This would streamline the regulatory process and facilitate regulatory harmonization. These two measures alone could significantly lower the cost of registration for diagnostics, reduce the delay to market entry and avoid duplication of in-country performance studies (McNerney, Sollis & Peeling, 2014).

Accelerating policy development

Most countries in the developing world do not have the capacity to develop robust diagnostic policies. Even when policies exist, the development is often very slow and not implementable because of the lack of resources in terms of both funding and health care personnel. And without the necessary policy in place, new and innovative diagnostic solutions may never enter the clinical pathway, where they are needed most. Again, building capacity for an HTA framework is a worthwhile investment as part of the AMR response. The framework would include the development of models to assess potential impact and cost–effectiveness of different strategies for deployment.

Novel financing mechanisms

In order to advocate the use of diagnostics to guide treatment decisions instead of the presumptive prescription of antibiotics for the common clinical syndromes (described in the Diagnostics for more targeted use of antibiotics section), financing mechanisms are needed for developing countries to procure diagnostics. Gavi, the Vaccine Alliance, is one example of such a mechanism. A diagnostic financing mechanism for low-resource settings which has been successful is the "buy-down" of tests by agencies such as the Global Fund to Fight AIDS, Tuberculosis and Malaria and UNITAID. This involves funding agencies that will procure the diagnostics from companies at volumes that allow substantially lower prices. It is not clear how sustainable this mechanism may be for countries, unless the countries come together to negotiate regionally. Surveys on test usage and volumes, and patient willingness to pay, would also allow companies to assess price points for the developed and developing world.

Educating the public on AMR

In emerging economies, educational campaigns to make the public and health providers aware of the importance of using a diagnostic test before treatment are critically important. Antibiotics are easily accessible and faster and less costly than a diagnostic test. Patients need to fully understand the long-term implications of inappropriate antibiotic use and antimicrobial resistance.

A more efficient system for implementation

Lessons learnt from existing POC tests should provide a starting-point for persuasive discussions on how to implement new diagnostics in a more efficient manner. Most countries need to develop plans and systems to support implementation. Understanding of the local contexts in which these technologies will be used is often overlooked (Boeras, Nkengasong & Peeling, 2017). Partners will need to come together to support the country plan. Connectivity solutions can be incorporated into laboratory systems managing a network of POC testing sites to create a more efficient system for training, supply chain management, quality assurance and monitoring safety and effectiveness (Cheng et al., 2016).

Return on investments

The impact of investments in novel technologies can only be realized with successful implementation and usage of quality diagnostics serving patient needs and public health. All the processes and systems that can bring this about should be measured to fully assess barriers and gaps to be addressed. Apart from promoting healthier lives, the most convincing arguments for countries to ensure that quality diagnostics are used to combat AMR would be to measure successes as returns on investment in lives saved and improved health outcomes.

Developing a business case for diagnostics for AMR

Traditionally, the business case for investing in a health product is made on the return on investment in terms of health benefits such as reduction in morbidity, the number of lives saved, transmissible infections averted, or costly long-term complications averted. This approach has worked well for advocacy for investment in drugs and vaccines (So et al., 2011). However, this approach has not worked well for making the business case for investments in diagnostics since many donors perceive that diagnostics, by themselves, do not save lives, compared to medicines and more direct interventions. Yet, it is widely acknowledged that diagnostics are important in disease control and prevention. The Lewin Report estimated that diagnostics account for less than 5% of health care costs but their results are used in 60–70% of health care decisions (The Lewin Group, 2005). Hence, a new approach is needed to advocate for the value of diagnostics in disease control and prevention, and in particular, for AMR.

This novel approach needs to model the contribution of diagnostics in reducing the threat of AMR in several aspects:

- Quantifying the risk of not having diagnostics to improve the specificity of syndromic management (i.e. maintaining the status quo for antibiotic prescriptions in primary health care and in hospitals).
- Assessing the impact of a new generation of connected diagnostics that can improve the efficiency of health care systems by simplifying patient pathways and guiding appropriate use of drugs and other resources.
- Developing models for investments in POC diagnostics that could be used to decrease the cost of drug trials through faster and more accurate means of identifying the target population for the drug trial.

Conclusion

Recent advances in POC technologies to ensure universal access to affordable quality-assured diagnostics have the potential to reduce misuse of antimicrobial compounds and improve patient outcomes. Innovation in diagnostics needs to continue to be stimulated by challenge prizes and supported through enabling structures such as access to biobanks with well-characterized specimens to facilitate test development. As technological innovation has steadily outpaced regulatory science, assessment of risks and benefit should no longer be done sequentially.

A new framework for HTA for joint review of risks and benefits by regulators and policy-makers, programme managers and subject matter experts is urgently needed, not only to facilitate a faster and more balanced regulatory review but also to accelerate implementation and policy development. Regional harmonization of a new HTA framework would also reduce duplication in clinical performance studies, reducing delays and lowering costs so that the marketed product becomes more affordable, and hence accessible.

For AMR surveillance to be effective, it is critical to: 1) understand the science and technologies needed for immediate pathogen identification to provide disease risk assessments and support global health decisions; 2) build a comprehensive network of laboratories and POC testing sites to implement quality-assured POC diagnostic services with a good laboratory–clinic interface; 3) use implementation science to understand the political, cultural, economic and behavioural context for novel diagnostic technology introduction.

As cost and funding will continue to affect innovations in diagnostics, a sound business case needs to be made to incentivize and de-risk R&D, and to finance novel diagnostic solutions for AMR. Quantifying the risk of not having diagnostics to improve the specificity of syndromic management can also encourage investment. In addition, it is important to assess the contribution of a new generation of connected diagnostics to improve the efficiency of health care systems by simplifying patient pathways, guiding appropriate use of drugs and other resources and improving patient outcomes.

References

Abdelmoez AT, Zaky DZ, Maher AM (2014). Role of cerebrospinal fluid IL-8 as a marker for differentiation between acute bacterial and aseptic meningitis. J Egypt Soc Parasitol. 44(1):205–210.

Boeras DI, Nkengasong JN, Peeling RW (2017). Implementation science: the laboratory as a command centre. Curr Opin HIV AIDS. 12(2):171–174.

Burdette AJ, Alvarez R (2014). Evaluation of innate immune biomarkers in saliva for diagnostic potential of bacterial and viral respiratory infection. Virginia: Defense Technical Information Center. (http://www.dtic.mil/docs/citations/ADA602373, accessed 06 September 2018).

Chalupa P, Beran O, Herwald H et al. (2011). Evaluation of potential biomarkers for the discrimination of bacterial and viral infections. Infection. 39(5):411–417.

Cheng B, Cunningham B, Boeras et al. (2016). Data connectivity: A critical tool for external quality assessment. Afr J Lab Med. 5(2):535.

Chin CD, Linder V, Sia SK (2012). Commercialization of microfluidic point-of-care diagnostic devices. Lab Chip. 12:2118–2134.

Derda R, Gitaka J, Klapperich CM et al. (2015). Enabling the development and deployment of next generation point-of-care diagnostics. PLoS Negl Trop Dis. 9(6):e0003857.

Dittrich S, Tadesse BT, Moussy F et al. (2016). Target product profile for a diagnostic assay to differentiate between bacterial and non-bacterial infections and reduce antimicrobial overuse in resource-limited settings: an expert consensus. PLoS One. 11(8):e0161721.

Ebrahim M, Gravel D, Thabet C et al. (2016). Antimicrobial use and antimicrobial resistance trends in Canada: 2014. Canada Comm Dis Rep. 42(11):227–231.

García PJ, Cárcamo CP, Chiappe M et al. (2013). Rapid syphilis tests as catalysts for health systems strengthening: A case study from Peru. PLoS One. 8(6)e66905.

García PJ, You P, Fridley G et al. (2015). Point-of-care diagnostics for low resource settings. Lancet Global Health. 3(5):257–258.

Global Antibiotic Research & Development Partnership (n.d.). Antimicrobial resistance. Landscape analysis of the state of AMR testing technologies. Geneva: Global Antibiotic Research & Development Partnership. (https://www.gardp.org/wp-content/uploads/2017/05/ AMR_Tech_Landscape_Analysis.pdf, accessed 06 September 2018).

Haran JP, Buglione-Corbett R, Lu S (2013). Cytokine markers as predictors of type of respiratory infection in patients during the influenza season. Am J Emerg Med. 31(5):816–821.

Horizon 2020 (n.d). Better use of antibiotics. Brussels: European Commission. (http://ec.europa.eu/research/horizonprize/index.cfm?prize=better-use-antibiotics, accessed 06 September 2018).

Huy NT, Thao NT, Diep DT et al. (2010). Cerebrospinal fluid lactate concentration to distinguish bacterial from aseptic meningitis: a systemic review and meta-analysis. Crit Care. 14(6):R240.

Ibrahim KA, Abdel-Wahab AA, Ibrahim AS (2011). Diagnostic value of serum procalcitonin levels in children with meningitis: a comparison with blood leukocyte count and C-reactive protein. J Pak Med Assoc. 61(4):346–351.

Jani IV, Peter TF (2013). How point-of-care testing could drive innovation in global health. N Engl J Med. 368(24):2319–2324.

Kapasi AJ, Dittrich S, González IJ et al. (2016). Host biomarkers for distinguishing bacterial from non-bacterial causes of acute febrile illness: A comprehensive review. PLoS One. 11(8)e0160278.

Kumar AA, Hennek JW, Smith BS et al. (2015). From the bench to the field in low-cost diagnostics. Angew Chem Int Ed Engl. 54(20):5836–5853.

Lee C, Walker SA, Daneman N et al. (2015). Point prevalence survey of antimicrobial utilization in a Canadian tertiary-care teaching hospital. J Epidemiol Glob Health. 5:143–150.

Linder A, Akesson P, Brink M et al. (2011). Heparin-binding protein: a diagnostic marker of acute bacterial meningitis. Crit Care Med. 39(4):812–817.

Lopman BA, Reacher MH, Van Duijnhoven Y et al. (2003). Viral gastroenteritis outbreaks in Europe 1995–2000. Emerg Infect Dis. 9(1):90–96.

Mabey DC, Sollis KA, Kelly HA et al. (2012). Point-of-care tests to strengthen health systems and save newborn lives: the case of syphilis. PLoS Med. 9(6)e1001233.

McNerney R, Peeling RW (2015). Regulatory in vitro diagnostics landscape in Africa: Update on regional activities. Clin Infect Dis. 61(Suppl 3): S135–140.

McNerney R, Sollis K, Peeling RW (2014). Improving access to new diagnostics through harmonised regulation: priorities for action. Africa J Lab Med. 3(1):A123.

Morel C, McClure L, Edwards S et al. (2016) Ensuring innovation in diagnostics for bacterial infection: Implications for policy. European Observatory Health Policy Series. Copenhagen: European Observatory on Health Systems and Policies. (http://www.euro.who.int-data/assets/pdf_file/0008/302489/Ensuring-innovation-diagnostics-bacterial-infection-en.pdf?ua=1, accessed 29 January 2019).

National Institutes of Health (2017). Antimicrobial resistance diagnostic challenge. Washington, DC: National Institutes of Health. (https://dpcpsi.nih.gov/AMRChallenge., accessed 06 September 2018).

Niemeier D, Gombachika H, Richards-Kortum R (2014). How to transform the practice of engineering to meet global health needs. Science. 345:1287–1290.

Nuutila J, Jalava-Karvinen P, Hohenthal U et al. (2013). A rapid flow cytometric method for distinguishing between febrile bacterial and viral infections. J Microbiol Methods. 92(1):64–72.

Okeke IN, Lamikanra A, Edelman R (1999). Socioeconomic and behavioral factors leading to acquired bacterial resistance to antibiotics in developing countries. Emerg Infect Dis. 5(1):18–27.

Okeke IN, Peeling RW, Goossens H et al. (2011). Diagnostics as essential tools for containing antibacterial resistance. Drug Resist Updat. 14:95–106.

O'Neill J (2016). Tackling drug-resistant infections globally: Final report and recommendations. The Review on Antimicrobial Resistance. London: Wellcome Trust and Government of the United Kingdom. (https://amr-review.org/sites/default/files/160518_Final%20paper_with%20 cover. pdf. accessed 06 September 2018).

Oved K, Cohen A, Boico O et al. (2015). A novel host-proteome signature for distinguishing between acute bacterial and viral infections. PLoS One. 10(3):e0120012.

Peeling RW, Nwaka S (2011). Drugs and diagnostic innovations to improve global health. Infect Dis Clin North Am. 25(3):693–705.

Rugera SP, McNerney R, Poon AK et al. (2014). Regulation of medical diagnostics and medical devices in the East African community partner states. BMC Health Service Research. 14:524.

Savuto P, Karuppan S (2017). Increased enrollment efficiency with point-of-care pre-screening. Applied Clinical Trials.26(8). (http://www .appliedclinicaltrialsonline.com/increased-enrollment-efficiency-point-care-pre-screening, accessed 06 September 2018).

So AD, Gupta N, Brahmachari SK et al. (2011). Towards new business models for R&D for novel antibiotics. Drug Resist Updat. 14(2):88–94.

Suarez NM, Bunsow E, Falsey AR et al. (2015). Superiority of transcriptional profiling over procalcitonin for distinguishing bacterial from viral lower respiratory tract infections in hospitalized adults. J Infect Dis. 212(2):213–222.

The Lewin Group (2005). The value of diagnostics innovation, adoption and diffusion into health care. Falls Church, VA: The Lewin Group. (https:// dx.advamed.org/sites/dx.advamed.org/files/resource/Lewin%20Value%20 of%20Diagnostics%20Report.pdf, accessed 06 September 2018).

The Longitude Prize (n.d.). Longitude Prize [website]. London: Challenge Prize Centre. (https://longitudeprize.org/, accessed 06 September 2018).

Turner K, Christensen H, Adams E et al. (2018). Analysis of the potential impact of a point-of-care test to distinguish gonorrhoea cases caused by antimicrobial-resistant and susceptible strains of Neisseria gonorrhoeae. London: Review on Antimicrobial Resistance. (https://amr-review.org/ sites/default/files/GCDiagnostictestmodel_WORKINGPAPER.pdf, accessed 06 September 2018).

UNITAID (2018). Fever diagnostic technology landscape, first edition. Geneva: UNITAID. (https://unitaid.eu/assets/Fever_diagnostic_technology_and_ market_landscape.pdf, accessed 06 September 2018).

University of Oxford (2015). Position paper on anti-microbial resistance diagnostics, June 2015. Oxford: Centre for Evidence-Based Medicine, Nuffield Department of Primary Care Health Sciences. (http://www.cebm .net/wp-content/uploads/2015/07/AMR-Diagnostic-technologies_10-June-2015.pdf, accessed 06 September 2018).

US Centers for Disease Control and Prevention (2013). Antibiotic resistance threats in the United States, 2013. Atlanta, Georgia: US Centers for Disease Control and Prevention. (https://www.cdc.gov/drugresistance/ threat-report-2013/index.html, accessed 06 September 2018).

Viallon A, Desseigne N, Marjollet O et al. (2011). Meningitis in adult patients with a negative direct cerebrospinal fluid examination: value of cytochemical markers for differential diagnosis. Crit Care.15(3):136.

Wang J, Wang P, Wang X et al. (2014). Use and prescription of antibiotics in primary health care settings in China. JAMA. 174(12):1914–1920.

Wi T, Lahra MM, Ndowa F et al. (2017). Antimicrobial resistance of Neisseria gonorrhoeae: Global surveillance and a call for international collaboration. PLoS Med. 14(7):e1002344.

World Health Organization (2015a). Global action plan in response to AMR. Geneva: World Health Organization. (http://www.who.int/ antimicrobial-resistance/publications/global-action-plan/en/, accessed 06 September 2018).

World Health Organization (2015b). Global antimicrobial resistance surveillance system manual for early implementation. Geneva: World Health Organization (http://apps.who.int/medicinedocs/documents/ s22228en/s22228en.pdf?ua=1, accessed 06 September 2018).

World Health Organization (2017a). Global framework for development and Stewardship to combat antimicrobial resistance. Geneva: World Health Organization. (http://www.who.int/phi/implementation/ research/ WHA_BackgroundPaper-AGlobalFrameworkDevelopment Stewardship-Version2.pdf?ua=1, accessed 06 September 2018).

World Health Organization (2017b). Global priority list of antibiotic-resistant bacteria to guide research, discovery, and development of new antibiotics. Geneva: World Health Organization. (http://www.who.int/medicines/ publications/WHO-PPL-Short_Summary_25Feb-ET_NM_WHO.pdf, accessed 06 September 2018).

Yager P, Domingo GJ, Gerdes J (2008). Point-of-care diagnostics for global health. Ann Rev Biomed Eng. 10:107–144.

Zaas AK, Burke T, Chen M et al. (2013). A host-based RT-PCR gene expression signature to identify acute respiratory viral infection. Sci Transl Med. 5(203):203ra126.

8 | The role of vaccines in combating antimicrobial resistance

MARK JIT, BEN COOPER

Introduction

The mitigation of global antimicrobial resistance (AMR) will require national and global coordination of interventions at the technological, behavioural, economic and political levels. Different tools need to be deployed intelligently; these include incentivizing new antimicrobial development, stewarding existing antimicrobials better, improving diagnostics to identify optimal treatments for patients and enhancing control measures to prevent infection spread (O'Neill, 2016). A key tool that offers unique advantages in combating AMR spread is the development and use of vaccines against infectious diseases.

The potential for vaccination as a tool against AMR has long been recognized, but has recently received renewed attention (Abbott, 2017). Reductions in antibiotic use and bacterial resistance were observed after the introduction of vaccines against *Streptococcus pneumoniae* (Fireman et al., 2003) and *Haemophilus influenzae* (Tandon & Gebski, 1991). Publicly commissioned reviews of AMR in the United Kingdom (O'Neill, 2016), the European Union (European Commission, 2017) and the United States (The White House, 2015) have pinpointed the role of vaccines as a key measure to lower demand for antimicrobials and hence combat AMR.

However, these reviews underestimate the potential benefit of vaccines because they only consider a subset of the pathways by which vaccines can affect antimicrobial use and resistance. Several additional reviews have recently outlined multiple interacting ecological, epidemiological and health systems pathways through which vaccines affect AMR (Lipsitch & Siber, 2016; Atkins et al., 2018).

How vaccines can reduce the burden of AMR

Vaccines used in humans

Vaccines may act to reduce the burden of AMR through a number of pathways (Lipsitch & Siber, 2016; Atkins et al., 2018). Below we briefly describe six potentially important effects to consider (Figure 8.1).

1) **Preventing infections by focal pathogens.** Vaccines may reduce the incidence of infection by a resistant pathogen. This can occur both through direct protection to those vaccinated, and through indirect protection resulting from reduced exposure to the infection in the unvaccinated (herd immunity). Use of vaccines that either reduce risk of infection or reduce transmission by infected vaccinees may result in these effects. A clear example of this is the *H. influenzae* type b (Hib) conjugate vaccine (see next section).

2) **Bystander effects.** Any vaccines that lead to changes in antibiotic use could potentially have an impact on AMR in organisms not targeted by the vaccine as a result of reduced antibiotic selection pressure. For example, since influenza infections are frequently treated with antibiotics (either inappropriately for the primary viral infection, or for a secondary bacterial infection), an effective and widely used vaccine that reduces the number of influenza infections should result in population-wide reductions in antibiotic use. In some cases, these reductions in antibiotic use may lead to reductions in resistance (Fireman et al., 2003; Kwong et al., 2008; Dagan et al., 2008). Since many bacterial pathogens are commonly carried asymptomatically and only rarely cause disease, such bystander effects may be the main path by which resistance is selected for. A reduction in bystander selection could thus have a substantial impact. It is important to note that the potential changes in antibiotic use resulting from vaccination may include both reductions in the volume of antibiotics used (fewer infections that need treating as a result of vaccination), as well as changes in the choice of antibiotic (for example, there may be reduced need for broad-spectrum antibiotics for a clinical syndrome if the vaccine reduces a resistant organism sufficiently that such cover is no longer required). Both changes could also lead to reductions in bystander selection of resistance.

3) **Infection severity effects.** Vaccines that reduce the risk of symptomatic infection without reducing the risk of carriage/asymptomatic infection can lead to reductions in the proportion of infections which are treated with antimicrobials and therefore a reduction in the selection pressure for resistant phenotypes. Malaria caused by *Plasmodium*

® denotes resistant pathogens. Antibiotic use may select for resistant pathogens; pathogens spread from person to person.

1. Preventing infections by focal pathogens. Vaccination may prevent resistance directly by preventing infection with the pathogen. Unvaccinated contacts of vaccinated people may also be protected due to herd immunity.

2. Bystander effects. Vaccination may lead to reduced antibiotic use due to fewer infections that are commonly treated with antibiotics. This may reduce selection for resistance in pathogens not targeted by the vaccine.

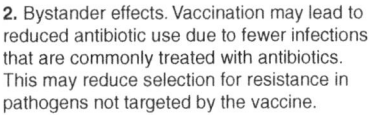

3. Infection severity effects. If vaccination reduces the risk of symptomatic infection, then even if it doesn't stop infection it can reduce antibiotic use, potentially reducing selection for resistance.

4. Subtype selection effects. If the vaccine protects against certain pathogen subtypes, and those subtypes tend to be more resistant, then resistance may be reduced.

5. Interspecific effects. Vaccinating against one pathogen may reduce risk of infection with another interacting pathogen of a different species.

6. Selective targeting effects. If resistant strains preferentially spread in certain populations, such as hospitals, then reducing transmission in such settings (e.g. through vaccines or monoclonal antibodies) will have a disproportionate effect on resistant strains.

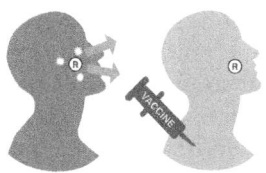

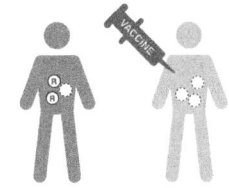

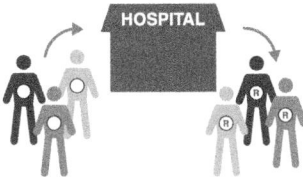

Figure 8.1 Ways in which vaccines may reduce antimicrobial resistance

Source: Authors' compilation.

falciparum provides an example where natural immunization may have played a role in slowing the emergence of resistance in some settings. In high malaria prevalence settings, levels of immunity in the population are higher than in low prevalence settings; as a result, a higher proportion of infections are asymptomatic and therefore untreated, leading to overall lower selection for resistance. This simple effect may explain why antimalarial resistance has historically arisen in lower-prevalence Asian settings rather than in higher prevalence areas of Africa (Pongtavornpinyo et al., 2008).

4) **Subtype selection effects.** When a pathogen population is composed of multiple competing subtypes and the vaccine targets only a subset of these, then if targeted subtypes are more likely to be resistant, overall resistance may decrease. The experience with pneumococcal conjugate vaccines provides an example of this (Dagan et al., 1996). Note that this type of selection could also lead to increases in resistance if subtypes targeted by the vaccine are less likely to be resistant, and initial reductions in resistance might be reduced by subsequent acquisition of resistance by non-vaccine serotypes (Mera et al., 2008).

5) **Interspecific effects.** Bacteria and viruses interact in complex ways. For example, influenza or respiratory syncytial virus (RSV) infections may increase the risk of secondary bacterial infections (Mera et al., 2008; McCullers, 2006; Bosch et al., 2013; Joseph, Togawa & Shindo, 2013) and patients with certain viral infections may transmit more bacterial pathogens (Eichenwald, 1960; Sherertz, 1996). Vaccination against one organism could therefore reduce transmission of another, leading to declines in both resistant and sensitive phenotypes. Note that in some cases competition effects between species could mean that vaccination against one organism could result in increases in risks of infection with competing species, as has been suggested for *Staphylococcus aureus* following pneumococcal vaccination (Brook & Gober, 2009).

6) **Selective targeting effects.** Interventions, such as hygiene improvements or vaccination, which might be expected to have equal effects on sensitive and resistant phenotypes of a given pathogen, could lead to differential effects if targeted to certain groups in structured populations. For example, if a resistant strain of a given pathogen transmits preferentially in hospitals (where antibiotic use is high), while a competing sensitive strains transmits better in the community, targeting the hospital population with a vaccine or passive immunotherapy conferring short-lived immunity would be expected to have

a greater overall effect on the resistant strain, leading to declines in resistance in both hospital and community (van Kleef et al., 2017).

Other, more complex, effects may also be in operation. For example, it is conceivable that vaccines could directly or indirectly alter bacterial populations in ways that lead to less or more opportunity for genetic exchange of resistance elements, potentially altering the rate at which resistance genes spread (Lipsitch & Siber, 2016).

It is also possible that vaccines will be developed to exploit other ways of reducing resistance. For example, it may be possible to develop vaccines to selectively target highly-resistant clones of a bacterial species. This may be possible by targeting toxins or other virulence factors which appear to be stably associated with resistant lineages (Wardenburg & Schneewind, 2008), or by targeting resistance determinants themselves (Senna et al., 2003; Zarantonelli et al., 2006; Joice & Lipsitch, 2013).

Vaccines used in animal production systems

The mechanisms by which vaccines may help reduce AMR in human pathogens also apply to animal production systems. Food-producing animals represent an important reservoir of drug-resistant bacteria and resistance genes. Such resistance may have detrimental effects on animal welfare and agricultural production. It may also affect human health through foodborne infections and through the transfer of resistance genes (Collignon et al., 2009). Antibiotic-resistant bacteria may spread from animals to humans either by direct contact or indirectly, through food preparation and consumption, via contaminated water, and through the use of animal waste as fertilizer (Marshall & Levy, 2011).

The World Organisation for Animal Health (OIE) has an ad hoc working group that focuses on the prioritization of diseases for which vaccines could reduce antimicrobial use in animals (World Organisation for Animal Health, 2015). This group is initially focusing on pigs, poultry and fish and aiming to identify the potential value of using both existing vaccines and potential new or improved vaccines targeting either bacteria where antimicrobial resistance is a problem or pathogens causing infections that are commonly treated with antibiotics in animal production systems. There may be even greater potential for reducing antimicrobial use in such animal production systems than there is in human populations, because of much greater rates of antibiotic use and the availability of vaccines which are not licensed for human use. This includes, for example, vaccines against *Escherichia coli* for which no corresponding vaccine is licensed for human populations, as well as many vaccines for pathogens which are important in animals but not in humans.

A good example of how vaccines have been able to reduce antimicrobial use in animal production systems is the experience of salmon farmers in Norway. Antibiotic use in commercial salmon farming in Norway was cut to virtually zero following development of a vaccine against furunculosis, a bacterial fish disease. Before the vaccine was introduced, large amounts of antibiotics were used, mixed in with the fish-feed. Following the introduction and widespread adoption of the vaccine this practice has been virtually eliminated (World Health Organization, 2015).

Advantages of vaccination

While vaccines are only one tool among many that can be used to combat AMR, they have a number of characteristics that make them particularly attractive for this purpose.

Vaccines work by enabling the immune system to recognize antigens that are highly specific to their targeted pathogens, and sometimes even to specific strains of a pathogen. In contrast, antibiotics work by targeting bacterial functions that are common across many species of microorganisms, both pathogenic and commensal. Consequently, vaccines usually have little or no effect on the evolution of microorganisms besides the targeted strains. In contrast, antibiotics can impose selective evolutionary pressure on both targeted and non-targeted microorganisms to develop resistance. Hence the development of resistance is likely to be simply a matter of time even if new generations of antibiotics are developed.

The specificity of vaccines opens up potential strategies that are unavailable with antimicrobials, let alone with nonpharmaceutical measures. For example, vaccines can be developed that target strains of a pathogen that are particularly pathogenic. The serotypes in the pneumococcal conjugate vaccines were generally selected to be the ones most likely to cause invasive disease (Hausdorff et al., 2000). This also means that they target the strains that are generally most likely to develop resistance, since pathogenic strains are the ones that lead to disease symptoms and thus antibiotic use. Going a step further, vaccines could be deliberately designed so that they target strains that are most likely to become resistant.

Furthermore, vaccines are usually deployed as a population-wide, preventive measure. Many vaccines are given in infancy and can protect their recipients for years or even decades. In contrast, antimicrobials need to be used sparingly because of their potential toxicity to the host and host microbiome, as well as to avoid the development of resistance. Hence, in most cases, they are administered only when people are already infected, and have to be continuously administered with each infection. This also means that they have less potential to prevent onward transmission of resistant microorganisms. Even if they reduce host infectivity, there is usually a delay between the onset of infectiousness and the time at which antimicrobials are received and become active. Hence vaccines offer greater potential for long-lasting population-wide effects that can prevent the onset of any disease at all. This potential is demonstrated by the eradication of smallpox and the near elimination of bacteria such as vaccine-targeted strains of *H. influenzae* and *Neisseria meningitidis* from countries with high coverage of the corresponding vaccines. Vaccines may be able to eradicate the underlying microbe rather than simply buy more time until resistance to a new antimicrobial emerges.

Vaccines and antimicrobials can work in a synergistic fashion – vaccines can reduce the rate at which populations are infected and hence extend the time until a pathogen evolves resistance to an antimicrobial. An antimicrobial can complement a vaccination programme by treating individuals who are unable to be vaccinated due to immune compromise.

Specific pathogens where vaccines and other immunotherapies could help reduce AMR burden

In this section we consider specific pathogens and the vaccines or other immunotherapies that could be used to combat AMR. We consider both currently licensed products and the pipeline for future immunotherapies. Information is summarized in Table 8.1, but key pathogens and associated vaccines are described in more detail below.

Streptococcus pneumoniae

The impact of vaccination with pneumococcal conjugate vaccines on the epidemiology of antibiotic resistance in *S. pneumoniae* has been

Table 8.1 *Vaccine and AMR status for selected important human pathogens*

Pathogen	Vaccine status	AMR status	Vaccine action on AMR
Bacteria			
Streptococcus pneumoniae	Conjugate vaccines targeting the main disease-causing pneumococcal serotypes are widely used. Vaccines with broad protection against all serotypes are being investigated	Common resistance to several classes of antibiotics, especially in vaccine serotypes	Direct and bystander effects
Haemophilus influenzae type b	Conjugate vaccine licensed and widely used	Widespread prior to vaccine introduction	Direct and bystander effects
Vibrio cholerae	Oral vaccine licensed and widely used	Growing single and multidrug resistance	Mostly direct
Salmonella typhi	Several vaccines available, with the new conjugate vaccine showing the most promising efficacy profile	Recent global expansion of resistant lineages which are associated with treatment failure to first-line antibiotics	Direct and bystander effects
Mycobacterium tuberculosis	Licensed widely used live vaccine (BCG) with variable, medium-term efficacy. Vaccines with higher and longer-lived protection in development	Growing single and multidrug resistance	Direct
Neisseria meningitidis	Vaccines against most of the main disease-causing serogroups (A, B, C, W and Y)	Some resistance to antibiotics	Direct
Clostridium difficile	Vaccines in development. Monoclonal antibodies licensed	*C. difficile* infection is associated with antibiotic use. Resistance to antibiotics in *C. difficile* itself has been observed.	Direct, bystander and selective targeting effects
Escherichia coli	Several vaccine candidates in development of Enterotoxigenic *E. coli* (ETEC) and	AMR a major concern for UPEC treatment	Direct and bystander effects

Group B *Streptococcus* (GBS)	No vaccines licensed, but several candidates are in development (capsular polysaccharide–protein conjugate vaccines and protein-based vaccines). Currently in phase I and II trials (Heath, 2016)	AMR is a concern primarily for patients who are allergic to first-line antibiotics. Resistance to second-line antibiotics is common in many settings (Castor et al., 2008)	Bystander effects might be expected to dominate given the high frequency of intrapartum antibiotic prophylaxis for GBS
Klebsiella pneumoniae	No vaccine candidates have been tested in humans	Multidrug resistance is endemic in many parts of world	
Neisseria gonorrhoeae	No vaccine at present, but indications that the vaccine against *N. meningitidis* type b may protect against *N. gonorrhoeae* (Petousis-Harris, 2017)	AMR is increasing and a serious concern in some cases leading to untreatable infections	

Viruses

Influenza virus	Several vaccines are licensed and widely used	Some resistance to antivirals	Bystander and interspecific effects
Rotavirus	Several vaccines are licensed and widely used	No antivirals in use	Bystander effects
Norovirus	Two vaccines have reached clinical trials. Several in preclinical stages	No antivirals in use	Bystander effects
Respiratory syncytial virus	Several vaccines are undergoing clinical trial	Not a concern	Bystander and interspecific effects
Dengue virus	Only licensed vaccine offers partial protection and may increase the risk of severe disease in some vaccinees	No antivirals in use	Bystander effects

Parasites

Plasmodium falciparum (malaria parasite)	Licensed vaccine offers partial protection	Some resistance to antimalarials	Direct

well-described and is one of the clearest examples of the potential impact of vaccination on AMR. In the United States the introduction of the seven-valent pneumococcal conjugate vaccine (PCV7) was associated with an 84% reduction in multidrug-resistant invasive pneumococcal disease (Kyaw et al., 2006). Reductions in resistance occurred through both direct effects (reduced total pneumococcal disease) and serotype selection effects (the vaccine serotypes were more likely to be resistant).

Haemophilus influenzae type b (Hib)

A conjugate vaccine against Hib is widely used. When it was first licensed, antibiotic resistance in Hib was an emerging problem, with both intrinsic resistance mechanisms limiting the activity of many antibiotics and highly-prevalent beta-lactamase production associated with resistance to ampicillin and amoxicillin (Tristram et al., 2007). The fact that this resistance is not of more concern is the result of the virtual elimination of serious infections with this pathogen (through both direct protection and herd immunity) in areas where population-wide Hib vaccination has been introduced. This highlights the potential of vaccines to reduce the AMR burden by simply reducing infections with the targeted pathogen.

Neisseria meningitidis

Resistance in the bacteria causing meningococcal disease has emerged, although it is not yet a widespread problem (Bash & Matthias, 2017). Vaccines now exist against most of the main disease-causing serogroups of *N. meningitidis* (A, B, C, W and Y).

Vibrio cholerae

Concerns about cholera, which is caused by the bacterial pathogen *V. cholerae,* have been greatly increased by the emergence of antibiotic-resistant lineages. Resistance to first-line antibiotics and multidrug-resistance both occur frequently and are associated with more severe illness (World Health Organization, 2017a; Gupta et al., 2016). Either one or two doses of the oral cholera vaccine are effective at preventing medically-attended cholera (Azman et al., 2016). Use of this vaccine clearly has the potential to reduce AMR through its direct effect on cholera.

Nosocomial bacterial pathogens

Multidrug resistance in bacterial species associated with nosocomial infection is common, facilitated by high levels of antibiotic use in health-care settings, high densities of patients, staff and visitors in close proximity as well as a high prevalence of immunocompromize among patients. Immunotherapies against several of these species are currently under development. Such therapies include active vaccines which lead to humoral and cellular immunity and passive immunotherapy which involves direct transfer of antibodies. Active vaccines can provide longer protection against target pathogens and are likely to be substantially cheaper. Among important nosocomial pathogens, pipelines for such active immunotherapies are limited to ongoing phase II/III clinical trials for *Clostridium difficile, Staphylococcus aureus* and *Pseudomonas aeruginosa*, with an earliest possible anticipated registration date of 2019 (Czaplewski et al., 2016).

Passive immunotherapies based on pathogen-specific monoclonal antibodies (mAbs), while only providing transient protection, have the potential benefit over active vaccines in that they confer immunity to the whole population including the elderly and immunocompromized who may often fail to develop effective immunity from active vaccines. Also, mAbs tend to have low rates of adverse reactions and can confer immunity in a shorter time after administration. The uses of mAbs are widespread as therapies for cancer and autoimmune diseases. While current costs represent a major barrier to wider adoption, production costs are expected to decrease as production volume increases and manufacturing technology matures. Notably, the Bill & Melinda Gates Foundation has recently invested more than $20 million in a biotech company, Achaogen, to accelerate the overall development of its mAbs discovery platform, in particular focusing on mAbs against the multidrug-resistant bacteria which are most problematic in low- and middle-income countries (LMICs) (Genetic Engineering & Biotechnology News, 2017). A number of mAbs are under development for *S. aureus* and *P. aeruginosa*, currently in phase I and II trials (Czaplewski et al., 2016). A monoclonal antibody against *C. difficile* toxin B, Bezlotoxumab, was approved by the US Food and Drug Administration in 2017 for use in the prevention of *C. difficile* infection (CDI) recurrence in patients under 18. This has been shown to reduce the risk of CDI recurrence from 26–28% to 16–17% (Wilcox et al., 2017). While *C. difficile* is not usually considered to be a pathogen where AMR is clinically important (as resistance does not impact on treatment), antibiotic resistance in *C. difficile* does occur and may significantly impact on the epidemiology and add to the burden of infection in settings where antibiotic use is high, as antibiotic-related disruption to the microbiome may provide an opportunity for *C. difficile* overgrowth. Notably, the large reductions in *C. difficile* infection that have been seen in England following multiple interventions have resulted primarily from reductions of fluoroquinolone-resistant strains (Dingle et al., 2017). Modelling work suggests that use of a vaccine or monoclonal antibodies targeted to health-care settings might be expected to have a substantially larger impact on resistant *C. difficile* strains if these spread preferentially in hospitals (van Kleef et al., 2017).

The "ESKAPE" pathogens *(Enterococcus faecium, Staphylococcus aureus, Klebsiella pneumoniae, Acinetobacter baumannii, Pseudomonas aeruginosa, and Enterobacter species)* are responsible for some of the most severe AMR problems. For example, in recent years the emergence of *Acinetobacter baumannii* lineages resistant to almost all antibiotics capable of treating Gram-negative bacteria has caused considerable alarm (Evans, Hamouda & Amyes, 2013). In some parts of the world the emergence of lineages resistant to first carbapenems (hitherto the last remaining conventional treatment option) and then colistin has severely compromised the ability to treat infections with this organism. Recently, multidrug-resistant *A. baumannii* has been estimated to account for over 36 000 deaths annually in Thailand (where carbapenem resistance is widespread), 41% more than if the infection had not been multidrug-resistant (Lim et al., 2016). This excess mortality makes this organism the dominant cause of multidrug-resistance-associated mortality in hospitalized patients in the country.

Similarly, multidrug-resistant *Klebsiella pneumoniae* has emerged in many healthcare settings as the leading cause of resistance-associated mortality and morbidity. For these reasons, carbapenem-resistant *A. baumannii*, and carbapenem-resistant, third-generation cephalosporin-resistant Enterobacteriaceae (which include *K. pneumoniae*), alongside carbapenem-resistant *P. aeruginosa*, were ranked in a recent World Health Organization (WHO) exercise as the pathogens of the highest priority where new antibiotics were needed – and, by extension, where the potential impact of vaccines could be greatest (World Health Organization, 2017b). Unfortunately, with the exception of *P. aeruginosa*, there is little activity in developing vaccines or other immunotherapies for these pathogens that has extended beyond animal models and it is thought unlikely that these vaccines will be available within the next 10 years. Major technical hurdles to developing vaccines for these "ESKAPE" pathogens exist and comprise limited understanding of pathogen biology including natural immunity, limited knowledge of vaccine targets, the existence of multiple strains and a complex and poorly defined epidemiology (Wellcome Trust and Boston Consulting Group, 2018). Lack of animal models of clear clinical relevance also presents an important hurdle for preclinical work, while the relatively low incidence of infections with these pathogens would make clinical trials of vaccines targeting these organisms challenging. Even if vaccines could be developed, the major difficulty in identifying a target population where vaccination would be cost-effective would present a high barrier to uptake.

Escherichia coli

Infections caused by *E. coli* are a major cause of morbidity and associated antibiotic use. In particular, enterotoxigenic *E. coli* (ETEC) is a leading cause of diarrhoea in children in developing countries and is estimated to account for 9% of all deaths attributed to diarrhoea (Lozano et al., 2012). Ciprofloxacin-resistant ETEC strains represent a major challenge for ETEC treatment strategies in some parts of the world (Begum et al., 2016). While there are no licensed vaccines for ETEC (Bourgeois et al., 2016), vaccine development for ETEC is a World Health Organization priority. WC-rBS, a killed whole-cell vaccine, designed and licensed primarily to prevent cholera, has been recommended by some groups to prevent diarrhoea caused by *E. coli*, although a Cochrane review found insufficient evidence to recommend it for protecting against ETEC diarrhoea (Ahmed et al., 2013). There are, however, a number of ETEC vaccine candidates in development and currently undergoing phase II trials, with one of the most advanced candidates a tetravalent inactivated whole-cell vaccine, ETVAX. While antibiotics are not recommended for the routine treatment of diarrhoea, even for infections such as ETEC which might respond (World Health Organization, 2005), in practice antibiotics are widely perceived as being the treatment of choice for diarrhoea in the community and were found to be used to treat about half of all episodes of diarrhoea in India and Kenya (Zwisler, Simpson & Moodley, 2013). The introduction of an ETEC vaccine could therefore potentially play an important role in reducing resistance primarily through its impact on reduced antibiotic consumption and therefore reduced bystander selection.

Urinary tract infections (UTIs) are the second most commonly seen infections in primary care (after respiratory tract infections). They account for a high volume of antibiotic usage in the community and represent a major economic and public health burden both of which are exacerbated by AMR (Flores-Mireles et al., 2015). The leading cause of uncomplicated community-acquired urinary tract infection is uropathogenic *E. coli* (UPEC). AMR in this organism is a major and increasing concern in the treatment of such infections, particularly in developing countries (Bryce et al., 2016). There has been considerable effort in the development of UPEC vaccines for the treatment of recurrent or chronic UTIs (Barber et al., 2013; Brumbaugh & Mobley, 2012). However, the fact that prior UTI does not elicit a protective immune response suggests this might not be easy. Another significant challenge is the high diversity of the UPEC population and multiple virulence factors (with no single one necessary for UTI).

Salmonella typhi

Antibiotic resistance has recently emerged as a major problem in *Salmonella typhi* infection, with the global dissemination of a ciprofloxacin-resistant lineage that is associated with fluoroquinolone treatment failure (Wong et al., 2015; Pham Thanh et al., 2016; Feasey et al., 2015). Two vaccines have been available since the 1990s and are recommended by the WHO: the live Ty21a vaccine and the Vi-polysaccharide vaccine (Anwar et al., 2014). Although both are effective in reducing typhoid fever, protection is partial and relatively short-lived, with Ty21 preventing one third to one half of typhoid cases in the first two years after vaccination and the Vi-polysaccharide vaccine preventing up to two thirds of cases in the first year after vaccination. Both vaccines have seen limited use in endemic countries and are mainly used to protect travellers to those countries. Several next-generation conjugate vaccines are in various stages of development, with two vaccine candidates having received licensure in India (Meiring et al., 2017). With Gavi, the Vaccine Alliance, having opened a funding window for this vaccine, it could potentially see greater use in LMICs.

Viral pathogens

Vaccines against viruses may play an important role in tackling AMR. This may occur both through reducing bystander selection by decreasing antibiotic usage for syndromes that may have either viral or bacterial causes and which are frequently treated with antibiotics (fever, respiratory infection, diarrhoea), and through reducing secondary bacterial infections that are causally linked to initial viral infections (such as influenza infection).

Among currently available viral vaccines, influenza vaccines (including inactivated influenza vaccines and live attenuated influenza vaccines) may have the greatest potential impact on AMR, with some observational data to support this. In 2000, the Canadian province of Ontario introduced universal influenza immunization. This was associated with a doubling of vaccine uptake to 28% (Kwong et al., 2009). This increase in uptake was associated with declines in rates of respiratory antibiotic prescriptions for most classes of antibiotics. Influenza-associated antibiotic prescriptions were estimated to represent 2.7% of respiratory antibiotic prescriptions before universal influenza vaccination, but only 1.1% of prescriptions afterwards. In Canadian provinces where influenza vaccination policy did not change, no such reductions in antibiotic prescribing were observed. While 2.7% seems a relatively small contribution to total antibiotic use (suggesting modest bystander selection effects and only a small benefit of vaccination), the fact that such antibiotic use is clustered in space and time could make selection effects for resistance locally more intense than population averaged figures suggest. Another study has estimated that reducing influenza activity by 20% would reduce fluoroquinolone prescriptions by 8%, although again translating such reductions to potential impact on resistance is challenging. We might expect impacts on AMR to be particularly large if children are targeted for vaccination. This is both because children typically consume more antibiotics than adults and also because they tend to play a disproportionate role in the transmission of both influenza and bacterial pathogens.

Similar benefits may occur if a vaccine against RSV can be deployed. Currently, there is no licensed RSV vaccine but a number are undergoing clinical trials, including vaccines that are likely to be optimal for the paediatric population as well as some that are likely to be appropriate for pregnant women and the elderly (Esposito & Di Pietro, 2016).

Vaccines against viruses causing diarrhoea may also lead to reductions in antibiotic use and reduced resistance by altering bystander selection. This may be important even though antibiotics are not usually recommended for treating diarrhoea, because, as noted above, in some parts of the world many caregivers still believe that antibiotics are effective at stopping diarrhoea, and inappropriate antibiotic treatment accounts for much of the costs of treating it (Zwisler et al., 2013). The currently licensed rotavirus vaccines may therefore play an important role, as they protect against the most common cause of severe diarrhoea in young children and prevent up to a third of severe diarrhoea cases in developing countries (World Health Organization, 2013; Soares-Weiser et al., 2012). A vaccine against norovirus would also be useful; norovirus accounts for nearly 20% of all cases of acute gastroenteritis which causes the second greatest burden of all infectious diseases globally (Ahmed et al., 2014). Currently, there is no licensed norovirus vaccine, but two candidate vaccines have reached clinical trials and there are a number of candidates at preclinical development stages (Cortes-Penfield et al., 2017).

Quantifying the economic benefit of vaccines that prevent antimicrobial resistance

The value for money of a vaccination programme can be estimated using an economic evaluation such as a cost–effectiveness analysis, which considers the balance between the incremental costs and incremental health impacts of an intervention. Hence, a vaccination programme will have a greater health impact and so will appear more cost-effective if it succeeds in reducing the absolute incidence of an antimicrobial-resistant pathogen and/or the prevalence of resistance in the pathogen. AMR reduction is likely to be a key benefit of vaccines such as those against *H. influenzae* serotype b (Bärnighausen et al., 2011). Yet a recent review of published models of the impact of vaccines on the dynamics of AMR did not find any studies that considered the economic value of this benefit (Atkins et al., 2018).

The benefit of vaccination in reducing AMR is normally seen in terms of avoiding the additional cost and increased health detriment incurred as a result of being infected by a resistant strain. This additional burden has been measured through matched cohort studies or regression models that compare the length of stay and treatment costs of patients infected with susceptible versus resistant microorganisms (Cohen et al., 2010). Hence, the simplest way to estimate the benefit of vaccination is to multiply the reduction in risk of acquiring a resistant strain in vaccinated individuals with the health detriment and financial cost of being infected with such a strain.

However, there are a number of ways in which this simple model of economic benefit may be too limited to comprehensively capture the value of a vaccine in reducing AMR. Firstly, public health interventions against infectious diseases may have wider ecological externalities on the community. For example, vaccines can confer protection not just on the people who have been vaccinated but also on other people who may otherwise have been infected by them (herd protection). Reduction of antimicrobial resistance through a number of pathways (see Introduction) is a further positive ecological externality of vaccines. Hence the benefit of vaccination should be measured not simply in individuals protected, but in the indirect effects in preventing both transmission and also reduction in AMR through several routes.

Second, several reviews of the economic benefits of vaccines have pointed out that economic analyses often overlook the wider benefits of vaccination on households and economies (Jit et al., 2015). This is particularly relevant in terms of prevention of AMR. The wider economic cost of AMR can be quite substantial when considering reductions in productivity from patients infected with resistant pathogens, the need to spend research and development funds in developing new antimicrobials because of resistance to existing antimicrobials, and the worst-case scenario of being unable to perform routine medical procedures such as surgery because of untreatable surgical site infections (Smith & Coast, 2013).

One further issue is that the scope of the positive externalities of vaccines in this regard may be global, since avoiding the development of resistance is a global benefit to all countries. However, the actors in the vaccine market are usually individual countries or even private individuals. Hence the full externalities of vaccination may not be adequately priced in a completely free market, discouraging manufacturers from developing vaccines that have benefits in reducing resistance. This market failure could be corrected by international cooperation and pooling of resources to encourage the development of vaccines with additional benefits in terms of AMR reduction. Mechanisms used in the vaccine world such as the Advanced Market Commitment for pneumococcal vaccines as well as pooled procurement by organizations such as Gavi and UNICEF offer models of this (De Roeck et al., 2006).

The importance of accurately capturing the value of vaccines against AMR

The vaccine development pipeline has been a busy one over the past few decades, and there are a number of vaccines that are currently being developed or trialled. As the number of vaccines in recommended schedules has increased, so has the cost of vaccination. For instance, the cost of fully vaccinating a child with all of the WHO's recommended vaccines has risen from $0.67 in 2001 to $45.59 in 2014 (Médecins Sans Frontières, 2015). As a result of the high cost of both procuring and delivering vaccines, vaccine access faces challenges particularly in low- and middle-income countries. For instance, there was a 12-year gap between the first high-income country introduction of hepatitis B vaccines and the first introduction in a low-income country (Gavi Alliance, 2012). Gavi was set up in 2000 as a public–private partnership in order to address this access gap. However, most middle-income countries, including some of the highest users of antibiotics, are not eligible (van Boeckel et al., 2014). Vaccine purchasers in middle-income countries need strong economic rationales to introduce new vaccines in the face of many competing health priorities. Hence, establishing the value proposition for vaccine development and use has become increasingly important.

Such introduction decisions are market signals for vaccine researchers and manufacturers to prioritize developing particular products and maintaining particular product lines over others (Robbins & Jacobson, 2015). Adequate incorporation of the full value of vaccines in preventing AMR will provide the right incentives for development of vaccines which target resistant organisms. This can also guide the decision over which groups to vaccinate. For example, pipeline vaccines against common hospital-acquired bacterial infections such as *S. aureus* and *C. difficile* may be prioritized for people most likely to receive antibiotics.

Most vaccines are developed for their main disease impact, with their impact on AMR a secondary consideration. However, properly valuing this secondary benefit is important in prioritizing the vaccine development pipeline, as well as choosing between alternative interventions (such as a new vaccine and a new antibiotic). Furthermore, precautionary vaccines may be developed against organisms such as *K. pneumoniae* that are currently of relatively minor health importance but may potentially emerge as a more serious threat should drug-resistant strains become widespread. Population models of the interplay between microbiological, ecological and economic forces in a population or even in the whole world are needed to quantify this additional benefit of vaccines and to see whether it would alter the priority ordering of vaccines that need to be developed. Methods for incorporating the impact on AMR into standard economic evaluations of vaccines were recently described, although it was also acknowledged that further research is needed to accurately quantify the pathways linking vaccination with health and economic outcomes (Sevilla et al., 2018).

Areas for future research

To fully capture the value of vaccines in reducing AMR, three sets of pathways need to be quantified:

1) The *health systems pathways* governing the impact of vaccines on antimicrobial prescriptions. These may be best informed by evidence from clinical trials as well as retrospective studies using electronic databases, such as those linking influenza vaccination with reduced antibiotic prescribing (see the Viral pathogens section).

2) The *epidemiological pathways* governing the impact of vaccines on AMR (both directly and through reduced prescribing). A major research question is how reductions in antimicrobial use following vaccination will (or will not) translate into reductions in resistance. In some cases there is clear evidence from observational data that reductions in certain combinations of antimicrobials are associated with decreased AMR combinations (Bell et al., 2014). However, this relationship is far from fully understood, and there is a long way to go before the impact that reductions in antimicrobial use will have on the prevalence of resistance could be reliably predicted. The task is complicated by uncertainty around the fitness cost of resistance (i.e. the ability of a resistant organism to reproduce compared to a susceptible one), since measurements in the laboratory may not translate into real world settings and compensatory mutations may mean such fitness costs change over time. There is also uncertainty around the importance of genetic hitchhiking (resistance levels may

rise not through any selective advantage, but as a result of fortuitous association with a successful and expanding lineage). From a modelling point of view, this step may require the use of dynamic transmission models that capture both direct and indirect effects of vaccines (Atkins et al., 2018), including herd effects through reduction of transmission, acquisition of resistant genes and competition between susceptible and resistant strains of pathogen.

3) The *economic pathways* governing the value of reduced AMR. Quantifying these pathways will ideally make use of models that capture the macroeconomic potential of vaccines reducing AMR over long-term time horizons, incorporating counterfactual scenarios such as the need to continuously develop new antibiotics and antibiotic classes, as well as the risk that entire medical procedures may become riskier or even impossible.

References

Abbott A (2017). Vaccines promoted as key to stamping out drug-resistant microbes. Nature. doi:10.1038/nature.2017.22324. (https://www.nature .com/news/vaccines-promoted-as-key-to-stamping-out-drug-resistant-microbes-1.22324, accessed 06 September 2018).

Ahmed SM, Bhuiyan TR, Zaman K et al. (2013). Vaccines for preventing enterotoxigenic Escherichia coli (ETEC) diarrhoea. Cochrane Database Syst Rev. 7:CD009029.

Ahmed SM, Hall AJ, Robinson AE et al. (2014). Global prevalence of norovirus in cases of gastroenteritis: A systematic review and meta-analysis. Lancet Infect Dis. 14(8):725–730.

Anwar E, Goldberg E, Fraser et al. (2014). Vaccines for preventing typhoid fever. Cochrane Database Syst Rev. 1:CD001261.

Atkins KE, Lafferty EI, Deeny SR et al. (2018). Use of mathematical modelling to assess the impact of vaccines on antibiotic resistance. Lancet Infect Dis. 18(6):e204–213.

Azman AS, Parker LA, Rumunu J et al. (2016). Effectiveness of one dose of oral cholera vaccine in response to an outbreak: A case-cohort study. Lancet Glob Health. 4(11):e856–63.

Barber AE, Norton JP, Spivak AM et al. (2013). Urinary tract infections: current and emerging management strategies. Clin Infect Dis. 57(5):719–724.

Bärnighausen T, Bloom DE, Canning D et al. (2011). Rethinking the benefits and costs of childhood vaccination: the example of the Haemophilus influenzae type b vaccine. Vaccine. 29(13):2371–23780.

Bash MC, Matthias KA (2017). Antibiotic resistance in Neisseria. Antimicrob Drug Res. 843–865.

Begum YA, Talukder KA, Azmi IJ et al. (2016). Resistance pattern and molecular characterization of enterotoxigenic Escherichia coli (ETEC) strains isolated in Bangladesh. PLoS One. 11(7):e0157415.

Bell BG, Schellevis F, Stobberingh E et al. (2014). A systematic review and meta-analysis of the effects of antibiotic consumption on antibiotic resistance. BMC Infect Dis. 14:13.

Bosch AA, Biesbroek G, Trzcinski K et al. (2013). Viral and bacterial interactions in the upper respiratory tract. PLoS Pathogens. 9(1):e1003057.

Bourgeois AL, Wierzba TF, Walker RI et al. (2016). Status of vaccine research and development for enterotoxigenic Escherichia coli. Vaccine. 34(26):2880–2886.

Brook I, Gober A (2009). Bacteriology of spontaneously draining acute otitis media in children before and after the introduction of pneumococcal vaccination. Pediatr Infect Dis J. 28(7):640–642.

Brumbaugh A, Mobley H (2012). Preventing urinary tract infection: Progress toward an effective Escherichia coli vaccine. Exp Rev Vaccines. 11(6):663–676.

Bryce A, Hay AD, Lane IF et al. (2016). Global prevalence of antibiotic resistance in paediatric urinary tract infections caused by Escherichia coli and association with routine use of antibiotics in primary care: Systematic review and meta-analysis. BMJ. 352:i939.

Castor ML, Whitney CG, Como-Sabetti K et al. (2008). Antibiotic resistance patterns in invasive Group B streptococcal isolates. Infect Dis Obstetr Gynecol. 2008:727505.

Cohen B, Larson EL, Stone PW et al. (2010). Factors associated with variation in estimates of the cost of resistant infections. Med Care. 48(9):767–775.

Collignon P, Conly JM, Andremont A et al. (2009). World Health Organization ranking of antimicrobials according to their importance in human medicine: A critical step for developing risk management strategies for the use of antimicrobials in food production animals. Clin Infect Dis. 49(1):132–141.

Cortes-Penfield NW, Ramani S, Estes MK et al. (2017). Prospects and challenges in the development of a Norovirus vaccine. Clin Ther. 39(8):1537–1549.

Czaplewski L, Bax R, Clokie M et al. (2016). Alternatives to antibiotics – a pipeline portfolio review. Lancet Infect Dis. 16(2):239–251.

Dagan R, Melamed R, Muallem M et al. (1996). Reduction of nasopharyngeal carriage of Pneumococci during the second year of life by a heptavalent conjugate pneumococcal vaccine. J Infect Dis. 174(6):1271–1278.

Dagan R, Barkai G, Givon-Lavi N et al. (2008). Seasonality of antibiotic-resistant Streptococcus pneumoniae that causes acute otitis media: A clue for an antibiotic-restriction policy? J Infect Dis. 197(8):1094–1102.

DeRoeck D, Bawazir SA, Carrasco P et al. (2006). Regional group purchasing of vaccines: review of the Pan American Health Organization EPI revolving fund and the Gulf Cooperation Council group purchasing program. Int J Health Plann Manage. 21(1):23–43.

Dingle KE, Didelot X, Quan TP et al. (2017). Effects of control interventions on Clostridium difficile infection in England: An observational study. Lancet Infect Dis. 17(4):411–421.

Eichenwald HF (1960). The "Cloud Baby": An example of bacterial-viral interaction. Arch Pediatr Adol Med. 100(2):161.

Esposito S, Di Pietro G (2016). Respiratory syncytial virus vaccines: An update on those in the immediate pipeline. Future Microbiol. 11:1479–1490.

European Commission (2017). A European One Health Action Plan against antimicrobial resistance (AMR). Brussels: European Commission. (https://ec.europa.eu/health/amr/sites/amr/files/amr_action_plan_2017_en.pdf, 06 September 2018).

Evans B, Hamouda A, Amyes S (2013). The rise of carbapenem-resistant Acinetobacter baumannii. Curr Pharma Des. 19(2):223–238.

Feasey NA, Gaskell K, Wong V et al. (2015). Rapid emergence of multidrug resistant, H58-lineage Salmonella typhi in Blantyre, Malawi. PLoS Negl Trop Dis. 9(4):e0003748.

Fireman B, Black SB, Shinefield HR et al. (2003). Impact of the Pneumococcal conjugate vaccine on otitis media. Pediatr Infect Dis J. 22(1):10–16.

Flores-Mireles AL, Walker JN, Caparon M et al. (2015). Urinary tract infections: epidemiology, mechanisms of infection and treatment options. Nature Rev Microbiol. 13(5):269–284.

Gavi Alliance (2012). Investing in immunisation through the Gavi Alliance – The evidence base. Geneva: Gavi Alliance. (https://www.gavi.org/ library/ publications/the-evidence-base/investing-in-immunisation-through-the-gavi-alliance---the-evidence-base/, accessed 06 September 2018).

Genetic Engineering & Biotechnology News (2017). Gates foundation supports Achaogen's antibacterial platform with $20.5M. New York: Genetic Engineering & Biotechnology News. (https://www.genengnews.com/ gen-news-highlights/gates-foundation-supports-achaogens-antibacterial-platform-with-205m/81254306, accessed 06 September 2018).

Gupta PK, Pant ND, Bhandari R et al. (2016). Cholera outbreak caused by drug resistant Vibrio cholerae serogroup O1 biotype ElTor serotype ogawa in Nepal: a cross-sectional study. Antimicrob Resist Infect Control. 5:23.

Hausdorff WP, Bryant J, Paradiso PR et al. (2000). Which pneumococcal serogroups cause the most invasive disease: implications for conjugate vaccine formulation and use, part I. Clin Infect Dis. 30(1):100–121.

Heath PT (2016). Status of vaccine research and development of vaccines for GBS. Vaccine. 34(26):2876–2879.

Jit M, Hutubessy R, Png ME et al. (2015). The broader economic impact of vaccination: reviewing and appraising the strength of evidence. BMC Medicine. 13:209.

Joice R, Lipsitch M (2013). Targeting imperfect vaccines against drug-resistance determinants: A strategy for countering the rise of drug resistance. PLoS One. 8(7):e68940.

Joseph C, Togawa Y, Shindo N (2013). Bacterial and viral infections associated with influenza. Influenza Other Respir Viruses. 7:105–113.

Kwong JC, Stukel TA, Lim J et al. (2008). The effect of universal influenza immunization on mortality and health care use. PLoS Med. 5(10):e211.

Kwong JC, Maaten S, Upshur RE et al. (2009). The effect of universal influenza immunization on antibiotic prescriptions: an ecological study. Clin Infect Dis. 49(5):750–756.

Kyaw MH, Lynfield R, Schaffner W et al. (2006). Effect of introduction of the pneumococcal conjugate vaccine on drug-resistant Streptococcus pneumoniae. N Engl J Med. 354(14):1455–1463.

Lim C, Takahashi E, Hongsuwan M et al. (2016). Epidemiology and burden of multidrug-resistant bacterial infection in a developing country. eLife. 5:e18082.

Lipsitch M, Siber G (2016). How can vaccines contribute to solving the antimicrobial resistance problem? mBio. 7(3):e00428.

Lozano R, Naghavi M, Foreman K et al. (2012). Global and regional mortality from 235 causes of death for 20 age groups in 1990 and 2010: A systematic analysis for the Global Burden of Disease Study 2010. Lancet. 380:2095–2128.

Marshall BM, Levy SB (2011). Food animals and antimicrobials: impacts on human health. Clin Microbiol. 24(4):718–33.

McCullers JA (2006). Insights into the Interaction between Influenza Virus and Pneumococcus. Clin Microbiol Rev. 19(3):571–582.

Médecins Sans Frontières (2015). The right shot: bringing down barriers to affordable and adapted vaccines, 2nd edn. Geneva: MSF (https://msfaccess .org/right-shot-bringing-down-barriers-affordable-andadapted-vaccines-2nd-ed-2015, accessed 06 September 2018).

Meiring JE, Gibani M, TyVAC Consortium Meeting Group (2017). The Typhoid Vaccine Acceleration Consortium (TyVAC): Vaccine effectiveness study designs: accelerating the introduction of typhoid conjugate vaccines and reducing the global burden of enteric fever. Report from a meeting held on 26–27 October 2016, Oxford, UK. Vaccine. 35(38):5081–5088.

Mera R, Miller LA, Fritsche TR et al. (2008). Serotype replacement and multiple resistance in Streptococcus pneumoniae after the introduction of the conjugate pneumococcal vaccine. Microb Drug Resist. 14(2):101–107.

O'Neill J (2016). Tackling drug-resistant infections globally: Final report and recommendations. The Review on Antimicrobial Resistance. London: Wellcome Trust and Government of the United Kingdom. (https://amr-review.org/sites/default/files/160518_Final%20paper_with%20 cover.pdf. accessed 06 September 2018).

Petousis-Harris H, Paynter J, Morgan J et al. (2017). Effectiveness of a group B outer membrane vesicle meningococcal vaccine against gonorrhoea in New Zealand: a retrospective case-control study. Lancet. 390:1603–1610.

Pham Thanh D, Karkey A, Dongol S et al. (2016). A novel ciprofloxacin-resistant subclade of H58 Salmonella typhi is associated with fluoroquinolone treatment failure. eLife 5:e14003.

Pongtavornpinyo W, Yeung S, Hastings IM et al. (2008). Spread of antimalarial drug resistance: Mathematical model with implications for ACT drug policies. Malar J. 7:229.

Robbins MJ, Jacobson SH (2015). Analytics for vaccine economics and pricing: insights and observations. Expert Rev Vaccines. 14(4):605–616.

Senna JP, Roth DM, Oliveira JS et al. (2003). Protective immune response against methicillin resistant Staphylococcus aureus in a murine model using a DNA vaccine approach. Vaccine. 21(19–20):2661–2666.

Sevilla JP, Bloom DE, Cadarette D et al. (2018). Toward economic evaluation of the value of vaccines and other health technologies in addressing AMR. Proc Natl Acad Sci USA. 115(51):12911–12919.

Sherertz R (1996). A cloud adult: The Staphylococcus aureus-virus interaction revisited. Annal Intern Med. 124(6):539–547.

Smith R, Coast J (2013). The true cost of antimicrobial resistance. BMJ. 346:f1493.

Soares-Weiser K, Maclehose H, Bergman H et al. (2012). Vaccines for preventing rotavirus diarrhoea: Vaccines in use. Cochrane Database Syst Rev. 2:CD008521.

Tandon MK, Gebski V (1991). A controlled trial of a killed Haemophilus influenzae vaccine for prevention of acute exacerbations of Chronic bronchitis. Aust NZJ Med. 21(4):427–432.

The White House (2015). National action plan for combating antibiotic-resistant bacteria. Atlanta. Georgia: Centres of Disease Control and Prevention. (https://

www.cdc.gov/drugresistance/pdf/national_action_plan_for_combating_
antibotic-resistant_bacteria.pdf, accessed 06 September 2018).

Tristram S, Jacobs MR, Appelbaum PC (2007). Antimicrobial resistance in
Haemophilus influenzae. Clin Microbiol Rev. 20(2):368–389.

van Boeckel TP, Gandra S, Ashok A et al. (2014). Global antibiotic consumption
2000 to 2010: an analysis of national pharmaceutical sales data. Lancet
Infect Dis. 14(8):742–750.

van Kleef E, Luangasanatip N, Bonten MJ et al. (2017). Why sensitive bacteria
are resistant to hospital infection control. Wellcome Open Res.2:16.

Wardenburg JB, Schneewind O (2008). Vaccine protection against
Staphylococcus aureus pneumonia. J Exper Med. 205(2):287–294.

Wellcome Trust and Boston Consulting Group (2018). Vaccines to tackle drug
resistant infections. An evaluation of R&D opportunities. Wellcome Trust
and Boston Consulting Group. (https://vaccinesforamr.org/wp-content/
uploads/2018/09/Vaccines_for_AMR.pdf, accessed 06 September 2018).

Wilcox MH, Gerding DN, Poxton IR et al. (2017). Bezlotoxumab for
prevention of recurrent Clostridium difficile infection. N Engl J Med.
376(4):305–317.

Wong VK, Baker S, Pickard DJ et al. (2015). Phylogeographical analysis of
the dominant multidrug-resistant H58 clade of Salmonella typhi identifies
inter- and intracontinental transmission events. Nat Genet. 47(6):632–639.

World Health Organization (2005). The treatment of diarrhoea. A manual for
physicians and other senior health workers. Geneva: World Health Organization.
(http://apps.who.int/iris/bitstream/10665/43209/1/9241593180.pdf,
accessed 06 September 2018).

World Health Organization (2013). Weekly epidemiological record. No.5,
2013, 88, 49–64. (http://www.who.int/wer/2013/wer8805.pdf?ua=1,
accessed 06 September 2018).

World Health Organization (2015). Vaccinating salmon: How Norway avoids
antibiotics in fish farming. Geneva: World Health Organization. (http://
www.who.int/features/2015/antibiotics-norway/en/.accessed 06 September
2018).

World Health Organization (2017a). Weekly epidemiological record.
No 34, 2017, 92, 477–500. (http://apps.who.int/iris/bitstream/
handle/10665/258763/WER9234.pdf;jsessionid=E86B2C79EB9363
0FB02C2B596EB1ED4A?sequence=1, accessed 06 September 2018).

World Health Organization (2017b). Global priority list of antibiotic-resistant
bacteria to guide research, discovery, and development of new antibiotics,
2017. Geneva: World Health Organization. (http://www.who.int/medicines/

publications/WHO-PPL-Short_ Summary_25Feb-ET_NM_WHO.pdf. accessed 06 September 2018).

World Organisation for Animal Health (2015). Report of the meeting of the OIE ad hoc group on prioritisation of diseases for which vaccines could reduce antimicrobial use in animals. Paris: World Organisation for Animal Health. (http://www.oie.int/fileadmin/SST/adhocreports/Diseases%20 for%20which%20Vaccines%20could%20reduce%20 Antimicrobial%20 Use/AN/AHG_AMUR_Vaccines_Apr2015.pdf, accessed 06 September 2018).

Zarantonelli ML, Antignac A, Lancellotti M et al. (2006). Immunogenicity of meningococcal PBP2 during natural infection and protective activity of anti-PBP2 antibodies against meningococcal bacteraemia in mice. J Antimicrob Chemother. 57(5):924–930.

Zwisler G, Simpson E, Moodley M (2013). Treatment of diarrhea in young children: Results from surveys on the perception and use of oral rehydration solutions, antibiotics, and other therapies in India and Kenya. J Glob Health. 3(1):010403.

9 | The role of civil society in tackling antimicrobial resistance*

ANTHONY D. SO, RESHMA RAMACHANDRAN†

The role of civil society: From public health to AMR

For decades, civil society has served as a critical catalyst in the public health arena. Civil society groups have played a role in moving policymakers and other stakeholders towards a new future, ensuring the right to access to essential medicines, and embracing the precautionary principle for environmental health risks. During the 1980s, organizations such as Health Action International (HAI) emerged onto the global policy scene, under the umbrella of Consumers International, to counter the pharmaceutical industry's promotion and pricing practices. The research and campaigns led by HAI and its allies resulted in increased public scrutiny of the marketing tactics used by the pharmaceutical industry for problem drugs, such as anabolic steroids used as appetite stimulants and vitamin tonics containing alcohol. Later, attention was drawn towards the impact of mark-ups on prices of medicines on their availability to populations worldwide (World Health Organization/Health Action International, 2008). This work contributed to international adoption of policies that ban direct-to-consumer advertising on prescription medicines. At the World Health Organization (WHO), these civil society organizations helped to shape the concept of the Essential Medicines List. At the country level, they also carried forward the WHO's Model Essential Medicines List by advocating and providing technical assistance towards the development and implementation of National Essential Medicines Lists, helping governments to secure affordable pharmaceutical prices. These initial efforts laid the groundwork for the access to medicines movement.

---

* The WHO, OECD and London School of Economics and Political Science do not endorse any commercial companies listed in any chapter throughout this book, and any policies presented are purely for academic purposes only.

† This chapter represents views, opinions and positions expressed in the authors' personal capacity, and these would not necessarily reflect the views of any third party, including the UN Interagency Coordination Group on Antimicrobial Resistance or the World Health Organization with which any of the authors might have affiliation.

The efforts made by civil society groups have increased public and policy-maker recognition of other health issues such as HIV/AIDS, non-communicable diseases, Ebola, tuberculosis, and tobacco use. Among these issues, civil society has played an active role in raising awareness on the necessity for rational use of antibiotics in reducing antimicrobial resistance (AMR) across the global north and south. During the 1980s, consumer organizations, such as HAI, the International Organization of Consumer Unions, the Medical Lobby for Appropriate Prescribing, and Oxfam, were already working on ending the promotion and marketing practices by multinational pharmaceutical companies, which targeted antibiotics among other pharmaceutical products in developing countries. These industry actions had contributed to the inappropriate use of these drugs and rising resistance (Kunin, 1993). Professional societies within the USA and Europe, such as the Infectious Diseases Society of America, the American Academy of Pediatrics, and Strama, developed updated antimicrobial treatment guidelines as well as stewardship programmes for physicians and other health care professionals to better conserve these life-saving drugs. In animal husbandry, Consumers International worked across countries to implement regulations around the nontherapeutic use of antimicrobials in food animal production and the rising levels of resistance in these products. In June 2000, Consumers International also provided a perspective at a WHO global consultation that focused on "general, overarching principles to reduce misuse and overuse of antimicrobials in animals intended for food" (World Health Organization, 2005).

In 1998, the WHO Member States already recognized AMR as a key global health issue and adopted a resolution at the 51st World Health Assembly that requested countries and the WHO to take action on research and development (R&D), access, and stewardship of antimicrobials across sectors (World Health Organization, 1998). In response to this mandate, the WHO put forward a Global Strategy for Containment of Antimicrobial Resistance in 2001 (World Health Organization, 2001). The Alliance for Prudent Use of Antibiotics had prepared an accompanying report compiling recommendations from groups around the world flagging the challenge of antibiotic resistance (Levy, 2010). However, the scheduling of the press conference to launch the release of this report could not have been more ill-fated as it coincided with the events of September 11, 2001, and thus never took place (Mack et al., 2011). Rekindling the WHO's return to this issue would become a priority for ReAct – Action on Antibiotic Resistance. Organized as a global policy network, the ReAct group formulated a Strategic Policy Program and met with WHO officials in support of the WHO's Patient Safety Programme, which examines AMR and patient safety. Their efforts were well received and supported by internal champions within the WHO, like Dr David Heymann, then a WHO Assistant Director-General. These key leaders then developed an international consultation process leading to the WHO monograph, *The evolving threat of antimicrobial resistance: options for action* (World Health Organization, 2012a). Margaret Chan, the WHO Director-General, announcing the release of the report, memorably described the threat of AMR: "A post-antibiotic era means, in effect, an end to modern medicine as we know it. Things as common as strep throat or a child's scratched knee could once again kill" (Chan, 2012).

The WHO's rekindled interest in AMR spurred others to follow, and global momentum quickly picked up pace with a number of key actions. The World Economic Forum highlighted antibiotic resistance in its *Global risks 2013* report (World Economic Forum, 2013). The World Health Assembly adopted a resolution in 2014, instructing the Secretariat to draft a Global Action Plan to combat antimicrobial resistance. Later the same year, the US President's Council of Advisors on Science and Technology released a report on antibiotic resistance, timed with an announcement from the White House of a National Strategy for Combating Antibiotic-Resistant Bacteria. Harnessing this global momentum, civil society groups continued their efforts to reset the policy-making process across sectors. Organizations such as Médecins sans Frontières (MSF) and HAI focused on ensuring equitable access to antimicrobials and preventive vaccines while others, such as the Alliance to Save Our Antibiotics and Food Animal Concerns Trust, advocated for regulations curbing the nontherapeutic use of antimicrobials in food animal production. However, this mobilization was disjointed, with organizations working within their own sectors of human and animal health.

Formation of the Antibiotic Resistance Coalition

Despite this, increased recognition of the One Health concept in AMR brought further awareness of the connections between using antimicrobials across human and animal health, as well as their impact on the environment. In March 2013, the Strategic Policy Program of ReAct – Action on Antibiotic Resistance proposed the creation of an intersectoral coalition of civil society groups that would tackle antibiotic resistance in collaboration with key organizations. This led to meetings between ReAct and several civil society organizations on how best to unify work on the human and animal use of antimicrobials. In so doing, these organizations considered how AMR policy concerns intersected with their priorities.

While recent policy declarations had signalled growing recognition of the challenge of antibiotic resistance, most avoided the political challenges of tackling the tougher issues: standing up to pharmaceutical industry calls for premium pricing, extending market exclusivity and efforts to lower drug regulatory and safety standards. Civil society groups also sought fair returns for public investment; conservation of existing antibiotics; and halting nontherapeutic use of antibiotics for not only growth promotion, but also routine preventive use in food animal production. These issues were shared concerns across civil society groups.

In the process, the ReAct Strategic Policy Program developed a systems framework (Figure 8.1) to provide a unifying framework to these discussions:

- access to life-saving antibiotics is a global concern, not just one of neglected diseases endemic in low- and middle-income countries (Access);
- the way antibiotic drugs are developed and brought to market influences how accessible these drugs will be (Innovation);
- the practices that govern antibiotic use in health care delivery affect how long these drugs can remain effective for use (Stewardship–Health Care Delivery);
- the use of antibiotics in the food production system, particularly for nontherapeutic purposes, poses risks of cross-species resistance (Stewardship–Food Production System); and
- antibiotics entering the environment, from wastewater discharge in manufacturing to point source pollution from hospitals and farms, indicates the need for an ecosystem approach to tackling antimicrobial resistance (Reimagining Resistance: Sustainability and Systems Thinking).

This systems framework became the foundational architecture of building buy-in, consensus towards the launch of a new coalition on antibiotic resistance, and the development of shared principles that comprised the Declaration of Shared Principles across these civil society groups (Figure 9.1).

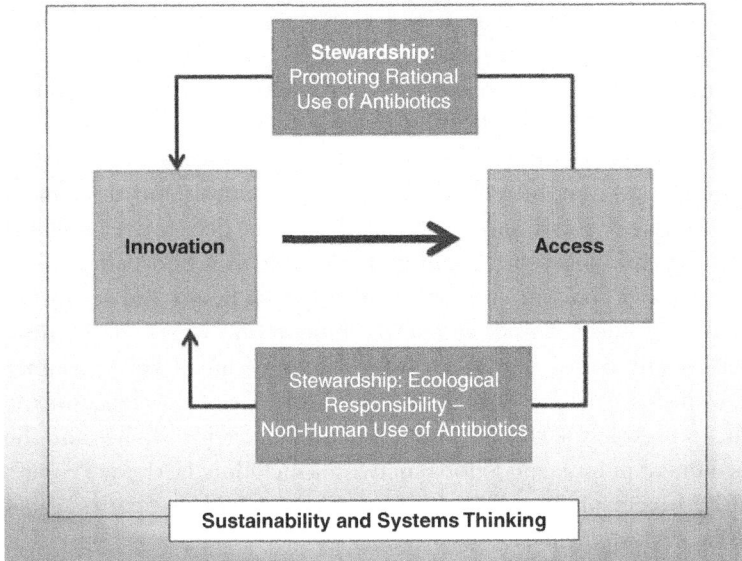

Figure 9.1 Systems diagram of the challenge of antimicrobial resistance

Source: So, 2014.

Building on a planning meeting around the World Health Summit in 2013, ReAct and a Steering Group of civil society organizations undertook a six-month process to lay the groundwork for a new coalition. The results of this process culminated in a conference hosted by the South Centre, an intergovernmental think tank for developing countries based in Geneva, and organized by the civil society Steering Group. The Steering Group worked to identify key civil society partners, common ground for collective concern and action, and a strategy for the launch of an intersectoral coalition and a Declaration of Shared Principles. In addition to the series of Steering Group teleconference calls, the ReAct Strategic Policy Program fielded an online consultative questionnaire of civil society groups to elicit early feedback. The founding meeting was held before the 2014 World Health Assembly where Member States would consider the adoption of a WHO resolution to develop a Global Action Plan on AMR.

The Geneva conference laid out the policy landscape, the challenges ahead, and importantly, cross-sectoral linkages. The conference agenda was designed to allow for group discussions over the five pillar areas highlighted in Figure 9.1. Each pillar corresponds to potential common ground – Innovation; Access but Not Excess; Human Use

of Antibiotics; Non-Human Use of Antibiotics; and Sustainability
and Systems Thinking. A drafting group emerging out of the Geneva
conference undertook the writing of the Declaration on Antibiotic
Resistance. The Declaration naturally coalesced around the pillars taken
up at the conference. The drafting group hammered out the consensus
over several weeks in the lead up to the World Health Assembly in
mid-May. HAI announced the finalized Declaration and the launch
of the Antibiotic Resistance Coalition during the debate over a World
Health Assembly resolution calling for a Global Action Plan against
AMR. Twenty civil society organizations from north and south, on
both sides of the Atlantic, and across human and veterinary sectors,
signed in support of the Declaration. These included key consumer
organizations, such as Public Citizen, the Center for Science in the
Public Interest in the United States, and the Centre for Science and the
Environment in India, and global networks including HAI, the People's
Health Movement, Third World Network, and the Universities Allied
for Essential Medicines.

Notably, the Declaration affirmed a shared set of key principles and
a commitment to safeguard the policy process from efforts that might
masquerade as solutions to tackling antibiotic resistance. Its principles
cut across sectors and call for:

- Realigning incentives in the health-care delivery system to support
  antibiotic stewardship;
- Curbing improper promotion and advertisement of antibiotics that
  might exacerbate inappropriate use of these drugs;
- Rethinking the metaphor of being at war with bacteria and instead
  to learn how to better live in harmony with the microbiome;
- Strengthening surveillance and transparency of antibiotic sales, use
  and resistance patterns;
- Eliminating the nontherapeutic use of antibiotics in producing food
  and encouraging the procurement of food products produced without
  nontherapeutic antibiotics by hospitals;
- Supporting incentives for pharmaceutical R&D for novel antibiotics
  and complementary technologies that delink a company's return on
  investment from volume-based sales;
- Opposing measures that undermine consumer safety by lowering clinical trial standards or place life-saving antibiotics out of
  affordable reach of those in need by extending monopoly pricing
  of drugs.

The Antibiotic Resistance Coalition (ARC) continues to be comprised of the original signatory organizations behind the Declaration. As part of the induction process into ARC, members are required to sign onto the Declaration on AMR principles and provide documentation of any potential financial conflicts of interest. A Nominating Committee of existing ARC members invites other aligned civil society organizations to join, and the ranks of the ARC have grown to include the Natural Resources Defense Council, the American Medical Student Association, MedAct, the Ecumenical Pharmaceutical Network, and the US Public Interest Research Group (PIRG). ARC serves as a platform for the member organizations to discuss key policy issues related to AMR, share organizational expertise across sectors and countries, and mount collective responses across policy forums, notably intergovernmental organizations such as the WHO and the UN.

Since its founding, the ARC has organized an annual WHO–NGO dialogue, a global teleconsultation that offers an opportunity for leading civil society groups to meet with key AMR leadership at the WHO to outline key concerns around ongoing policy processes. The dialogues are held strategically in advance of the World Health Assembly. These discussions allow civil society organizations, including those in developing countries, to engage directly with policy-makers on key upcoming decisions. In April 2015, ARC held its first WHO–NGO dialogue with the WHO Assistant Director-General Keiji Fukuda and AMR Coordinator Charles Penn. This dialogue focused on the draft Global Action Plan on AMR in advance of its discussion and anticipated adoption at the World Health Assembly in May 2015. Here, civil society put forward interventions on technical and financial support for implementing the Global Action Plan, the challenges of innovation, access, and rational use of antimicrobials, the intersectoral concerns over the use of antimicrobials in agriculture, and how trade treaties influence the use of these drugs in food products, and the need for accountability, monitoring, and evaluation. Subsequent WHO–NGO dialogues, held again in advance of the World Health Assembly, have touched on the implementation of the Global Action Plan including the global development and stewardship framework, the WHO's role in supporting the creation and implementation of national action plans, and the UN High-Level Meeting on Antimicrobial Resistance held in September 2016. Each year, full summaries of the teleconsultation are published, providing a record and roadmap of the interventions made by ARC members and allies (Antibiotic Resistance Coalition, 2017).

As AMR discussions unfolded at the United Nations, the ARC has played a key role in advocating for the coalition principles to be reflected within the Political Declaration from the UN High-Level Meeting on Antimicrobial Resistance. Leading up to the negotiations around the Political Declaration, ReAct co-hosted a UN briefing on AMR along with the Dag Hammarskjold Foundation and the UN Secretary-General's "Every Woman Every Child" initiative. Here, ARC members and civil society allies including ReAct, Food Animal Concerns Trust, and MSF delivered interventions outlining specific points around innovation, access, and stewardship across sectors to Member States and UN agencies in order to influence upcoming negotiations around the Political Declaration on AMR. The findings of this discussion called for the final UN Political Declaration on AMR to ensure broader interagency accountability beyond the tripartite collaboration of the WHO, the Food and Agricultural Organization (FAO), and the World Organisation for Animal Health (OIE) to the UN level. Additionally, in advance of the UN High-Level Meeting on AMR, ARC members and allies met with country missions in New York and Geneva to call for parity between human and animal health. These points were reflected in the final document adopted during the High-Level Meeting on AMR at the UN General Assembly in September 2016.

Innovation

The need to bring new antibiotics to market gave momentum to growing policy-maker concerns over drug resistance. Civil society has played a key role in triggering this policy concern by documenting the dearth of novel antibiotics in the R&D pipeline, rekindling this discussion at the WHO and in other key forums, and connecting this to larger concerns over innovation and access to essential medicines. The lens of antibiotic resistance presented an opportunity to revisit policy issues from a different vantage point. *Access*, but not *excess* meant striking the right balance in stewardship of these resources. This would also require aligning the economics with the biology of drug resistance. The traditional business model of drug companies earning returns on investment from volume-based sales fails to do this. Moreover, the need for life-saving antibiotics is not limited to low- and middle-income countries, which places these issues beyond the exceptionalism or special regard argued for neglected diseases that primarily affect the world's poorest populations.

Almost all novel classes of antibiotics that were brought to market in recent decades were discovered before the 1990s. This faltering R&D pipeline became the focus of civil society attention. An early study focused on the shortfall of antibiotic drug candidates in the pipeline of multinational drug firms (Spellberg et al., 2004). Going further, ReAct partnered with the European Medicines Agency and the European Centre for Disease Prevention and Control to produce an analysis that examined all known antibacterial drug candidates in the R&D pipeline. Among these candidates, the study found not a single drug with a novel mechanism of action targeting Gram-negative pathogens (Freire-Moran et al., 2011).

This evidence supported the Swedish European Union (EU) conference, "Innovative Incentives for Effective Antibacterials" in 2009 and the establishment of the Transatlantic Task Force on Antimicrobial Resistance (TATFAR) with the United States. With Swedish government support, ReAct convened in 2010 an international conference "The Global Need for Effective Antibiotics: Moving Towards Concerted Action" to follow up. The conference notably brought existing public–private partnerships together to discuss "Reengineering the Value Chain for Research and Development of Antibiotics: Applying Lessons from Neglected Diseases". To facilitate this discussion, the conference featured a panel that included the TB Alliance, India's Open Source Drug Discovery Initiative, and the Drugs for Neglected Diseases Initiative (DNDi). Prior to the conference, ReAct's policy team held discussions with the Director-General of the Swedish pharmaceutical industry trade association, Richard Bergström (later the Director-General of the European Federation of Pharmaceutical Industries and Associations). At the conference, Bergström acknowledged in a report (Braine, 2011) that:

> [i]ncentives that separate the financial return from the use of a product are the only way to change this behavior. Intelligent pull incentives, such as advance commitments and prizes, provide financial rewards to the developer that are not based on the volume of use of the novel antibiotic.

Following the conference, proceedings focused on new business models for R&D of novel antibiotics and echoed this conclusion to delink a company's return on R&D invested in a drug from its volume-based sales (So et al., 2011).

The concept of delinkage has its roots in debates about ensuring access to medicines. Delinkage represented an approach, advanced by civil society, that promised fairer drug pricing and returns on public investments in R&D. This is typically accomplished by divorcing the drug company's return on investment from R&D from the price of the drug. Back in 2004, Jamie Love of Knowledge Ecology International and Tim Hubbard (Hubbard & Love, 2004) envisaged that countries might commit a small percentage of their gross domestic product to global health R&D in exchange for lifting Trade-Related Aspects of Intellectual Property Rights requirements on World Trade Organization (WTO) members to comply with patent protections that blocked the market entry of generic medicines.

This concept of delinkage became a key principle in the WHO's Consultative Expert Working Group on Research and Development report in 2012 (World Health Organization, 2012b). With respect to antibiotic innovation, delinkage also has to separate the return on R&D investment from volume-based sales, or in other words, the price and quantity of antibiotics sold (So & Shah, 2014). Increasingly, delinkage has entered policy discussions on both sides of the Atlantic, from Chatham House to the US President's Council of Advisors on Science and Technology (Clift et al., 2015; US President's Council of Advisors on Science and Technology, 2014). Civil society also actively supported its inclusion in the WHO's Global Development and Stewardship Framework on AMR and the UN Political Declaration on AMR.

The call for greater support of drug development came not only from the pharmaceutical industry, but also from the public sector. The UK Review on AMR proposed that $16 billion would be required to reinvigorate the R&D pipeline, assuming that 15 new antibiotics – including four breakthrough drugs – would come to market over the next decade (O'Neill, 2016). The Boston Consulting Group's report for the German Ministry of Health recommended that the investment for each commercialized product would amount to $1 billion, plus $200 million per year for a Global Research Fund to develop the infrastructure for developing promising projects, and $200 million annually for a Global Development Fund to support all stages of clinical development (Stern et al., 2017).

By January 2016, the Davos Declaration by Pharmaceutical, Biotechnology and Diagnostics Industries on Combating Antimicrobial Resistance signalled the industry's commitment for "appropriate incentives (coupled with safeguards to support antibiotic conservation) for companies to invest in R&D", "pricing of antibiotics [that] more adequately reflects the benefits they bring", and "novel payment models that reduce the link between the profitability of an antibiotic and the volume sold" (International Federation of Pharmaceutical Manufacturers & Associations, 2016a). Later that year, a subgroup of these companies developed an "Industry Roadmap for Progress on Combating Antimicrobial Resistance". In this roadmap, the industry noted that the "receipt of an adequate Market Entry Reward will greatly facilitate global access and stewardship for that product" and "progress incentives, such as lump-sum payments, insurance models and novel IP [intellectual property] mechanisms, that reflect the societal value of new antibiotics and vaccines and will attract further investment in R&D" (International Federation of Pharmaceutical Manufacturers & Associations, 2016b). However, the Roadmap fails to mention delinkage as such.

By contrast, MSF has called for full delinkage between a company's return on R&D investment from price and volume of the drug sold (Sanjuan, 2017). ReAct has not only advanced the concept of delinkage, but also has questioned whether the emphasis on market entry rewards fails to address adequately the key scientific bottleneck in the R&D pipeline (So et al., 2017). ReAct's Strategic Policy Program put forward proposals for collaborative R&D approaches, two of which Regional WHO Offices advanced to the top 22 proposals for global consideration as part of the WHO's Health R&D Demonstration projects. One proposal focused on building a diagnostic innovation platform to address antibiotic resistance, while the other concerned establishing a drug discovery platform for sourcing novel classes of antibiotics as public goods. These civil society positions on AMR derive from their previous advocacy on access to medicines for treatment of HIV/AIDS, tuberculosis, malaria and other neglected diseases.

However, AMR also moved policy discussions beyond the exceptionalism of neglected diseases. Product development partnerships (PDPs) had focused on neglected diseases, the treatment of which posed little competition to industrialized country markets. These PDPs had successfully recruited in-kind contributions from industry; however, such approaches were considered part of the exceptionalism of non-paying markets. By contrast, AMR affects patients everywhere in the world, and new approaches to innovation in this area could not be viewed as exceptionalism. Civil society's efforts brought to the fore what the industry was slow in acknowledging – 30 years of a faltering antibiotic R&D pipeline demanded public sector intervention.

The WHO laid important groundwork for a public–private partnership to support antibiotic innovation. A series of policy discussions with stakeholders would lead to the launch of the Global Antibiotic Research and Development Partnership (GARDP) as a project within the DNDi. Civil society played an important role, from advancing a range of potential proposals to supporting the WHO's own concept of a publicly-financed global consortium to tackle antibiotic resistance (World Health Organization, n.d.). Notably among PDPs, DNDi, which had received start-up funding from MSF, has worked to build access and capacity in countries where these most neglected diseases are endemic, and has included key research institutions and government ministries from low- and middle-income countries on its governance board. GARDP, in the spirit of DNDi's previous work, has also held consultations with civil society groups as it has begun to chart its course in developing new antibiotics.

Tackling AMR means more than simply bringing new drugs to market or making existing ones more available to those in need. It involves decreasing the selective pressure on existing antibiotics through improved diagnostics and vaccines. Civil society has actively worked on both. MSF and ReAct worked with the Foundation for Innovative New Diagnostics, a product development partnership focused on diagnostics, and the WHO to bring experts together to discuss biomarkers that might distinguish bacterial from other infectious causes of acute fever (World Health Organization et al., 2015). The MSF Access campaign waged a global effort to lower the price of the pneumococcal conjugate vaccine, manufactured by Pfizer and GlaxoSmithKline (GSK). MSF's "A Fair Shot" campaign argued that with one million children dying each year from pneumonia, Pfizer's and GSK's pricing of the vaccine limited the possible reach of this potentially life-saving intervention (Médecins sans Frontières, n.d.) (Figure 9.2). In fact, if universal coverage with pneumococcal conjugate vaccine had been achieved in the 75 countries where vaccination rates fell short of 80%, nearly half the days of antimicrobial therapy to treat children less than 5 years old for pneumonia could have been averted (Laxminarayan et al., 2016) (Figure 9.2).

HERE'S WHY YOU SHOULD CARE

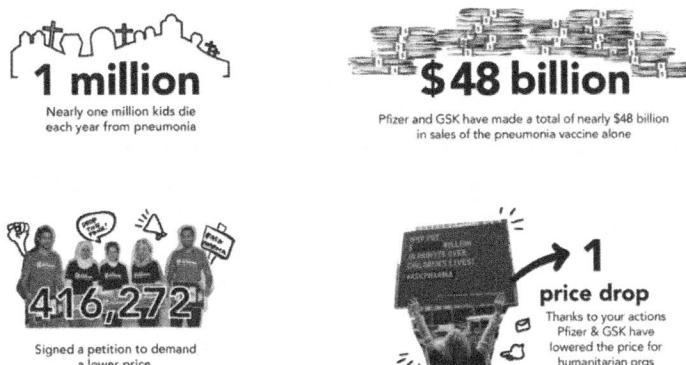

Figure 9.2 "A Fair Shot" pictograph by the Médecins sans Frontières Access Campaign

Source: Médecins sans Frontières, n.d.

Access not excess

As proposals to invigorate the antibiotic pipeline began to emerge, questions loomed large on how access as well as stewardship of these drugs might be achieved. Civil society had been advancing the idea of full delinkage. However, other proposals were put forth and ran counter to these full delinkage models. These partial delinkage models would still apply close-to-marginal cost pricing and controls over quantity in low- and middle-income countries, but would not apply the same to industrialized country markets. Civil society had opposed such proposals in forums, ranging from DRIVE-AB to the UK Review on AMR (ReAct, 2017). In their analysis of the final recommendations from the UK Review on AMR, MSF expressed concern that market entry rewards were only seen as a way to delink volume, but not price of the product. MSF opposed this reframing of delinkage as a tool that ensures stewardship, but did not address affordable access (Médecins sans Frontières, 2016).

In the USA, consumer groups, including ARC members, and allies, including Public Citizen and MSF, have squared off with industry, the Infectious Diseases Society of America, and the Pew Charitable Trusts over a number of incentive proposals introduced as legislation in Congress. In response to the Generating Antibiotic Incentives Now (GAIN) Act, which awards extended data exclusivity to newly approved antibiotics, consumer groups noted how such monopoly protections give companies an incentive to sell more of the new drug. Instead of providing upfront investments in R&D, such incentives just risk imposing higher drug prices on consumers (So & Weissman, 2012). Rationing antibiotics by monopoly pricing will not ensure appropriate use by doctors or patients.

These groups and others, including professional societies, also expressed concern over proposals, such as the 21st Century Cures Act, to lower regulatory standards for approval of new antimicrobials. The 21st Century Cures Act weakened the Food and Drug Administration's (FDA) drug regulatory protections by replacing gold standard reliance on clinical trials with "adaptive" pathways and surrogate end-points. In a post in the *Health Affairs Blog,* members of the National Physicians Alliance FDA Task Force noted that lowering regulatory standards would incentivize the development of more expensive, me-too drugs of "marginal or ultimately insignificant effectiveness" (Molchan et al., 2015). Civil society has also expressed concern over proposals for transferrable exclusivity extensions allowing manufacturers facing patent expiry to acquire additional monopoly price protections (Alas, 2017; Seabury & Sood, 2017).

Besides countering proposals that would hinder affordable access to novel antimicrobials or other complementary technologies, civil society has advocated for a set of core principles established in the access to medicines movement and the Antibiotic Resistance Declaration. These core principles include delinkage, affordability, availability, effectiveness and quality. Civil society has carried forward these principles – initially adopted as part of the recommendations of the WHO's Consultative Expert Working Group (CEWG) on R&D – to other intergovernmental policy forums in an effort to create coherence around these processes. In November 2015, the UN Secretary-General announced the creation of the High-Level Panel on Access to Medicines with the mandate of examining proposals and recommending solutions that would address the policy incoherence between inventors and trade rules, on the one hand, and international human rights law and public health, on the other. Seeing this as an opportunity to shape the language of global governance and demonstrate an alternative vision for the future, civil society quickly became activated.

Civil society put forward over half of the almost 200 contributions towards the High-Level Panel's recommendations, while representatives from organizations such as MSF, Oxfam, the Health Global Access Project and Lawyers Collective served as part of the Expert Advisory Group to the High-Level Panel. In the panel's final report, AMR was highlighted as a case-study with the recommendation that innovation models applying delinkage be pursued as a way to ensure sustainable access to novel antimicrobials. This inclusion of AMR as a specific case study and the accompanying call for delinkage, rather than market-based models, again demonstrates how the issue has become an item on the global health policy agenda because of support from civil society groups.

As part of implementing the Global Action Plan on AMR, the WHO Director-General was mandated at the 68th World Health Assembly (2015) to develop options for a global development and stewardship framework on AMR. As the WHO and its partners within the tripartite collaboration began to develop this framework, civil society called for the CEWG principles, including full delinkage from both price and quantity, to be reflected in the policy documents. ARC members and allies including the South Centre, Third World Network, MSF, and ReAct urged Member States and the WHO to safeguard access to antimicrobials and other complementary technologies such as vaccines and diagnostics. As the consultative process has continued, these principles have been incorporated into key policy documents reflecting civil society's success in shaping the policy language on these points. Through continued efforts by civil society, these principles were also incorporated into the Political Declaration adopted at the UN High-Level Meeting on AMR.

As reports of rising resistance to last-line antimicrobials continued to emerge around the world, so did the urgency to ensure stewardship to preserve the effectiveness of these life-saving drugs for those in need. The conservation of these drugs must also be balanced by the need for appropriate access – access, but not excess. The lack of access to antibiotics remained a serious concern, particularly in developing countries. Treatable infectious diseases are estimated to claim the lives of 5.7 million people a year (Daulaire et al., 2015). Additionally, as civil society has pointed out, three quarters of deaths from community-acquired bacterial pneumonia could be averted if antibiotics were universally available to children under 5 years old (Laxminarayan et al., 2016). This lack of access, however, is not just from shortages or stockouts of these medicines, but also from drug resistance rendering these antimicrobials ineffective. Resistance to first-line antibiotics has been estimated to result in over 56 000 neonatal deaths in India and over 25 000 neonatal deaths in Pakistan (Laxminarayan & Bhutta, 2016). According to UNICEF, pneumonia and diarrhoea account for more than one out of every four children dying under the age of five. Yet, fewer than a third of children with suspected pneumonia received antibiotics. Additionally, while fewer than four in 10 children receive treatment with oral rehydration for diarrhoea, they instead receive inappropriate treatment with antibiotics (UNICEF, 2016). The key is to ensure access, but not excess.

"Access but not excess" became an important refrain advanced by civil society, from its contribution to the Lancet Infectious Diseases Commission to the WHO–NGO dialogue discussions. Concerns over underuse, not just overuse, parallel the public statements made by low- and middle-income delegations such as India and Brazil. At the 70th World Health Assembly in May 2017, Dr Lav Agarwal of the Permanent Mission of India noted that India objects to any "unbalanced emphasis" on a Stewardship Framework focused on limiting access to antibiotics as opposed to R&D and affordable access to new and existing antibiotics and diagnostics (Agarwal, 2017). In October 2016 at the WHO/WIPO/WTO Joint Technical Symposium on AMR, Dr Lucas Vinícius Sversut of the Permanent Mission of Brazil stressed that "avoiding unnecessarily restrictive policies is particularly important for developing countries, where the lack of access to antimicrobial medicines kills more than the resistance itself" (Sversut, 2016).

Enlisting health care professionals in antimicrobial stewardship is critical. In the United States, Health Care Without Harm and the PIRG have also engaged health professionals around both human and animal use of antimicrobials. Working across hospitals, Health Care Without Harm has developed a number of procurement guidelines for purchasing meat and seafood products raised without the nontherapeutic use of antibiotics (Health Care Without Harm, 2015). Going further, regional Healthy Food in Health Care programmes were established to allow for collaborative efforts across hospitals and institutions locally to boost the market demand for meat raised without routine antibiotics. The organization also formed the Clinician Champions in Comprehensive Antibiotic Stewardship (CCCAS) Collaborative as an initiative to raise awareness among health-care professionals on the link between antibiotic use in agriculture and AMR. Along with increased awareness, this should contribute towards the promotion of policies for the judicious use of these therapies across sectors (Health Care Without Harm, 2015). In collaboration with the Pediatric Infectious Disease Society and Sharing Antimicrobial Reports for Pediatric Stewardship group, Health Care Without Harm provides tools for clinicians to take local action at their home institutions to change their purchasing practices. CCCAS members are also trained to relay their professional experiences on the impact AMR has had on their patients and public health as a way to promote policy action for stewardship. PIRG has also mobilized over 40 000 health care and public health professionals through its Health Professionals Action Network to call on major restaurant chains to source meat products raised without the routine use of antibiotics and to adopt public policies with this commitment (US Public Interest Research Group, n.d.). Through this network, clinicians are also given the opportunity and support to voice their experiences with AMR to policy-makers.

Non-human use of antibiotics

Unlike most other areas of access to medicines, antimicrobial resistance has a One Health dimension. Although challenging to quantify, a significant proportion of all antibiotics, by volume, are sold for use in agriculture and aquaculture. In the US, this figure approaches 70% (US Food and Drug Administration, 2015). This situation has created an unusual convergence of interests across civil society movements. Traditionally, groups working on nutrition, the environment, animal welfare and worker justice have focused on the food system. Their work involves a quite different set of stakeholders – agribusiness concerns and those more focused on the FAO and OIE.

The use of antibiotics to enhance productivity in food animal production goes back decades. In the interval, livestock production has undergone growing intensification, reliant on practices requiring greater antibiotic use. Between 2010 and 2030, antimicrobial consumption in food animal production is predicted to rise by 67%. Two thirds of this increase can be traced to the increase of animals in food production, and a third, to the shift towards more intensive farming operations (Van Boeckel et al., 2015). Antimicrobials have an appropriate role in treating diseased animals, but a significant part of what is used for food animal production is nontherapeutic. Civil society groups working to curb the unnecessary use of antibiotics in food animal production consider therapeutic use as treatment when there is a diagnosis of disease. The use of antibiotics for growth promotion – to reduce losses in production or increase weight gain – would thus be considered nontherapeutic. In Europe, bans on the use of antimicrobials for growth promotion in food animal production in 2006 did not curb the sales of antimicrobials critically important for human medicine. Only when further measures were taken to restrict the routine preventive use of these antibiotics, as in Denmark, did antibiotic use decrease.

For these groups, nontherapeutic use extends to routine preventive use of antibiotics. In a report by the Alliance to Save Our Antibiotics (2016), Cóilín Nunan observed that:

> The shocking overuse of farm antibiotics shown by these data is a result of the continued failure by most countries to ban routine preventative mass medication in intensive farming. Spain now uses 100 times more antibiotics per unit of livestock than Norway, 80 times more than Iceland and 35 times more than Sweden. The main reason for the difference is that Spain, like most of Europe, allows routine mass medication, whereas the Nordic countries do not.

He further notes that to meet the UK Review on AMR's target of 50 mg of antibiotic per kilogram of livestock, Europe would need 65 years to achieve this goal starting at 152 mg/kg and reducing use at the current rate of 2% per year.

In the United States, the Food and Drug Administration (FDA) brought therapeutic uses of antibiotics in food animal production under the supervision of veterinarians. It also successfully obtained agreement from the veterinary drug manufacturers producing medically important antibiotics to remove voluntarily indications on their products for growth promotion or improving feed efficiency. This approach could readily be implemented in a country where there are only 26 manufacturers (Food and Drug Administration, 2013). However, Keep Antibiotics Working, a coalition of US-based groups, argued that such measures fell short because it did not ban the use of antibiotics for routine disease prevention (Keep Antibiotics Working, 2014). The FDA acknowledged civil society's "concern that drug manufacturers may promote extra-label production uses for products approved only for therapeutic use, thereby undermining the spirit and intent of [agency guidance]" (Hopkinson, 2014). In the lead up to the implementation of this agreement between veterinary drug manufacturers and the FDA, Keep Antibiotics Working pointed to examples of advertising from Novartis and Elanco who were still touting the use of their drugs for growth promotion (Zuraw, 2014).

Despite the fact that over 10% of the world's antibiotics in food animal production are used in the US, policy-makers in that country have moved much more slowly than policy-makers elsewhere to curb nontherapeutic use of antibiotics. Failing to make significant headway in changing US government policy, key civil society groups took up a different strategy. These groups included the Consumers Union, PIRG, the Center for Food Safety, Friends of the Earth, the Natural Resources Defense Council (NRDC) and the Food Animals Concerns Trust. Together, they called upon the country's largest restaurant chains to source their food from animal products produced without the routine use of antibiotics. Their demands were threefold (NRDC, 2015):

1) Immediate action to end the routine use of antibiotics important for human medicine.
2) A time-bound action plan to phase out any routine use of antibiotics across the supply chain.
3) The adoption of third-party auditing and verification of compliance with the antibiotics use policy, implementing and bench-marking results to show progress in meeting the goals described above.

The campaign has reached a larger scale in recent years. Targeting *Yum! Brands*, the conglomerate owner of KFC, Pizza Hut and Taco Bell, PIRG canvassers went door-to-door, gathering over 475 000 petition signatures. This generated thousands of calls to KFC customer service lines. PIRG's "KFC Save ABX" campaign resulted in hundreds of social media actions directed at the company, particularly by young people whom KFC had been targeting to rebrand its appeal.

Targeting Subway's employees, NRDC also commissioned a billboard outside the company headquarters that read "Is Subway Buying Meat Produced with Antibiotics?" A few days later, the company added to its website:

> Our commitment to serve high quality, affordable food to our customers has always been a cornerstone of the SUBWAY brand. We support the elimination of sub-therapeutic use of antibiotics. Elimination will take time and we continue to work with our suppliers to reach that goal (Brook, 2015).

As with public-facing brands, these restaurant chains have proved sensitive to consumer pressure and have begun to respond. The consumer groups created a public scorecard grading the top 25 companies on the US market. In each of the first three years, the *Chain Reaction* report has registered significant gains. Fourteen of these companies have begun to address limiting antibiotics in their supply chain. These initial steps have largely occurred in the poultry supply chain. Commitments to remove routine use of antibiotics in pork and beef supplies have lagged behind. Despite this, Chipotle and Panera have led with exemplary policies while Subway plans to address pork and beef, but on a much longer timeline (Friends of the Earth, 2017). Most of these commitments though are limited to US restaurants and franchises. However, in August 2017, McDonald's announced that it would meet its goal of serving broiler chicken not treated with antibiotics a year ahead of schedule in the USA. Updating its "Vision for Antimicrobial Stewardship for Food Animals", McDonald's plans to extend this commitment to eliminate the use of highest priority critically important antibiotics from its poultry supply chain globally. It will carry this out stepwise between 2017 and 2027 (McDonald's, 2017).

Other groups have targeted procurement efforts at different points in the supply chain. For example, Health Care without Harm's "Healthy Food in Health Care" programme draws upon the purchasing power of health-care institutions to advance sustainable food system practices. Partnering with over 1 000 hospitals across North America, Healthcare without Harm has worked to shape procurement policies in the healt care sector to support goals, such as sourcing food animal products raised without the routine use of antibiotics. School Food FOCUS and the Pew Charitable Trusts developed the Certified Responsible Antibiotic Use standard for chicken sold to institutional purchasers, which disallows the use of "antibiotics with analogues in human medicine routinely or without clear medical justification" and requires third-party certification to audit the supply chain for compliance (USDA, n.d.; Antibiotic Resistance Action Center, 2016). NRDC supported the Urban School Food Alliance, which includes six of the largest city school districts in the USA (New York City, Dallas, Orlando, Chicago, Los Angeles, and Miami-Dade), in its efforts to adopt the Certified Responsible Antibiotic Use Chicken policy. New York City's school system ranks as one of the country's largest institutional providers of meals, second only after the Department of Defense, and serves 860 000 meals per day.

Similar efforts can also be seen in non-western countries. In the Republic of Korea, the Ministry of Agriculture, Food, Forestry & Fisheries began to phase out the routine use of antibiotics in commercial compound feed in 2003 (USDA Foreign Agricultural Service, 2011). After initially reducing the number of antibiotics permitted in commercial compound feed from 53 to 25 in 2005, the process continued until all remaining antibiotics were removed by 2011. At a UN briefing on AMR, co-organized by the UN Secretary-General's "Every Woman Every Child" Initiative, the ReAct Network and the Dag Hammarskjold Foundation, Yong-Sang Kim (Director of Animal Health Management Division for the Ministry) acknowledged the important role of consumer groups in supporting these policy changes (Figure 9.3).

As the global dialogue has unfolded on AMR, the WHO's work with its sister intergovernmental agencies, notably the FAO and OIE, has grown. Codex Alimentarius, whose work is supported as part of the Joint FAO/WHO Food Standards Programme, has also received attention from

Figure 9.3 Dutch Minister of Health, Welfare and Sport, Edith Schippers, poses for photo with US Public Interest Research Group at the 2016 UN General Assembly

Source: Austin Donohue, US Public Interest Research Group (PIRG), September 2016.

the civil society as it examines the standards, guidelines and codes of practice affecting the use of antimicrobials in food production. Codex's role as the key organization in setting trade rules for food safety as an organization is recognized by the World Trade Organization's Sanitary and Phytosanitary Agreement. Several of those engaged in civil society actions on AMR in the USA have represented consumer interests on these issues before WHO expert committees, such as the Advisory Group on Integrated Surveillance of Antimicrobial Resistance, or as part of Consumers International's delegation before the Codex Alimentarius Commission. The Antibiotic Resistance Coalition has also channelled civil society concerns into the public consultation process held by the UN Interagency Coordination Group on AMR and by the Tripartite Monitoring and Evaluation framework on indicators to benchmark progress on AMR. Of particular importance, ARC has sought to be inclusive of civil society groups from low- and middle-income countries.

Sustainability and systems thinking

From its outset, the Antibiotic Resistance Coalition recognized the importance of sustainability and systems thinking in civil society's work on AMR. It was one of the pillars of the Antibiotic Resistance Declaration, and all corners of the coalition approach sustainability from differing vantage points – economic, environmental and cultural.

Civil society has focused on how to ensure the long-term sustainable access to antibiotics. This requires ensuring fair returns on R&D investment, affordable pricing of antibiotics, and effective stewardship such that these products can have lasting value in human medicine. From an economic vantage point, the concept of delinkage seeks to unify these three goals. The emergence of a product development partnership committed to such goals could be a game changer in how new models of innovation become piloted.

In addition to human and animal health, the third part of the One Health triangle addresses the role and impact of AMR on the environment. A growing number of publications document wastewater contamination with antibiotics, beginning with the manufacturing plants producing active pharmaceutical ingredients of these life-saving drugs. The antibiotic effluent from these plants has reached levels toxic to local life forms, but has also resulted in inducing drug-resistant pathogens in the environment. A series of reports from the Changing Markets

Foundation exposed these polluting practices among Indian and Chinese drug manufacturing plants, where much of the world's drug production occurs. Amplifying a recommendation from the UK Review on AMR, these reports also called for those buying generic antibiotics to consider the environmental track record of the manufacturers from which the drugs are sourced in making procurement decisions (Changing Markets et al., 2016).

Under the umbrella of the Antibiotic Resistance Coalition, other civil society groups have discussed how to address the point sources of antibiotic pollution into the environment. The motivations of industry efforts to reduce pollution from generic manufacturing plants in India and China were also considered. The question has also been raised as to why the same industry groups are not equally concerned about the likely far greater point source pollution posed by antibiotic use in agricultural run-off and hospital waste discharge. Are some multinational firms looking for an advantage in a competitive market, or are they truly concerned about the environmental contamination posed by the production and use of antibiotic drugs?

The 2016 Davos "Declaration by the Pharmaceutical, Biotechnology and Diagnostics Industries on Combating Antimicrobial Resistance", signed by over 100 companies, makes a brief mention of support for measures to curb antibiotic effluents into the environment (IFPMA, 2016a). Later that year, a far smaller number of companies signed the "Industry Roadmap for Progress on Combating Antimicrobial Resistance" (IFPMA, 2016b). The industry roadmap calls for several measures to mitigate the environmental impact from the production of antibiotics, but remained silent on the environmental impact from use of antibiotics. Most antibiotic residues are clearly not discharged from manufacturing plants, but rather from hospitals and farms.

Tackling this environmental discharge, Healthcare without Harm has begun to examine the waste management practices of hospitals. In lieu of medical waste incineration, the organization has identified no-burn technologies as part of an inventory of safer solutions (Emmanuel & Stringer, 2007). The Centre for Science and Environment (CSE) in Delhi has looked more broadly at how to integrate animal and environmental aspects into the development of National Action Plans on AMR in developing countries. With participants from 18 countries, the CSE organized a workshop that shared challenges and best practices in addressing surveillance and responsible use of antibiotics in food animal production and in the environment (Centre for Science and Environment, 2017).

Going further, the environment can also shape the response to the challenge of AMR. From its founding in 2005, ReAct has sought to go beyond the "war metaphor" in addressing AMR. Not long after, an Institute of Medicine report, *Ending the War Metaphor: The Changing Agenda for Unraveling the Host–Microbe Relationship*, described the rising tide of new pathogens and the need for a new paradigm, one that:

> incorporates a more realistic and detailed picture of the dynamic interactions among and between host organisms and their diverse populations of microbes, only a fraction of which act as pathogens ... The time has come to abandon notions that put host against microbe in favor of an ecological view that recognizes the interdependence of hosts with their microbial flora and fauna and the importance of each for the other's survival. Such a paradigm shift would advance efforts to domesticate and subvert potential pathogens and to explore and exploit the vast potential of nonpathogenic microbial communities to improve health (Forum on Microbial Threats, Board on Global Health, and Institute of Medicine, 2006).

Taking a cultural approach, ReAct Latin America has rooted their call for holistic solutions that address the interconnected relationship of bacteria and humans in the indigenous peoples' concept of *sumak kawsay*. An ancient Quechua phrase, *sumak kawsay* refers to "good living" or the "good life", living in harmony with ourselves, our community, and nature. Closely aligned to these efforts, ReAct also supported the *Microbes and Metaphors* project, in which a dialogue among scientists, artists and activists took place. Those involved in the project have raised important questions about the shortcomings of the biomedical paradigm. The editors of a volume of their collected works argue:

> One of the main reasons for this lack of progress in dealing with the phenomenon of "resistance" seems to be the flawed "war metaphor" which shapes the way antibiotics are used against pathogenic bacteria ... even more fundamentally we need to ask whether it is productive at all to constantly frame questions about the microbial world in an anthropocentric manner without considering the breathtaking diversity and even aesthetic beauty of the microbial world? (Sivaraman & Murray, 2015).

Figure 9.4 Book on microbes by children for children from ReAct Latin America

Source: ReAct, n.d.

From this corner of civil society, they draw inspiration from artists and ask "Can artists, sensitive to the ecological processes that govern all life forms, help us frame our questions in a better manner or gather new insights where our stale words fail us?" (Figure 9.4).

Conclusion

Just as civil society catalysed global attention over monopoly pricing of patented HIV/AIDS drugs, new civil society actors have redirected attention from rational use to the dearth of novel antibiotics in the R&D pipeline. Rekindling attention to AMR at the WHO contributed to the policy momentum that brought the issue to the world stage. This was supported by a Global Action Plan and a UN Political Declaration. AMR, by its nature, demands an intersectoral response. This gave impetus to efforts to create an intersectoral alliance, the Antibiotic Resistance Coalition, which brought together a number of civil society groups unified by shared principles. Civil society organizations have successfully introduced the concept of delinkage into the policy vernacular and mobilized consumer pressure on major restaurant chains to source food animal products raised without routine use of antibiotics. This work is remarkable because of the complexity of the AMR issue, its intersectoral nature, and the fact that its victims do not readily identify themselves with this shared global health challenge. While ReAct's vision of ensuring a future free from the fear of untreatable infections is years away, the remarkable richness of the contributions that civil society has made to the policy discussions and debates over AMR offers a useful compass for future policy-making.

References

Agarwal L (2017). Response from India delivered by Sh. Lav Agarwal, Joint Secretary, Ministry of Health on issue of Antimicrobial resistance. World Health Assembly, 70. (http://pmindiaun.org/pages. php?id=1478, accessed 06 September 2018).

Alas M (2017). Civil Society and South Centre call for urgent actions to tackle AMR and ensure innovation models. Geneva: South Centre News. (https://www.southcentre.int/south-centre-news-on-amr-1-9- august-2017/, accessed 06 September 2018).

Alliance to Save Our Antibiotics (2016). Massive overuse of farm antibiotics continue in Europe. Bristol: Sustainable Food Trust. (http:// sustainablefoodtrust.org/articles/massive-overuse-farm-antibiotics continues-europe/, accessed 06 September 2018).

Antibiotic Resistance Action Center (2016). School Food FOCUS Certified Responsible Antibiotic Use standard (CRAU). Washington, DC: George Washington University's Milken Institute School of Public Health. (http:// battlesuperbugs.com/sites/battlesuperbugs.com/files/ CRAU%20Rationale%20 and%20Standard%20July%202017.pdf, accessed 06 September 2018).

Antibiotic Resistance Coalition (2017). 3rd Annual WHO-NGO dialogue report. Antibiotic Resistance Coalition (http://abrcoalition.com/3rd-annual-who-ngo-dialogue-report-april-2017/, accessed 06 September 2018).

BIO Intelligence Service (2013). Study on the environmental risks of medicinal products. Final report prepared for the executive agency for health and consumers. Paris: BIO Intelligence Service. (https://ec.europa.eu/health/ sites/health/files/files/environment/study_environment.pdf, accessed 06 September 2018).

Braine T (2011). Race against time to develop new antibiotics. Bull World Health Organ. 89(2):88–89.

Brook L (2015). Subway moves the needle on antibiotics, but many questions remain. NRDC Expert Blog, 1 September 2015. (https://www.nrdc.org/experts/lena-brook/subway-moves-needle-antibiotics-many-questions-remain, accessed 06 September 2018).

Centre for Science and Environment (2017). Strategic and operational guidance on animal and environmental aspects: national action plans on antimicrobial resistance for developing countries. New Delhi: Centre for Science and Environment. (https://cdn.cseindia. org/userfiles/strategic-and-operational-guidance-final.pdf, accessed 06 September 2018).

Chan M (2012). Antimicrobial resistance in the European Union and the world. Geneva: World Health Organization. (http://www.who.int/dg/speeches/2012/amr_20120314/en/, accessed 06 September 2018).

Clift C, Outterson K, Rottingen J-A et al. (2015). Towards a new global business model for antibiotics: delinking revenues from sales. London: Chatham House. (https://www.chathamhouse.org/publication/towards-new-global-business-model-antibiotics-delinking-revenuessales, accessed 06 September 2018).

Daulaire N, Bang A, Tomson G et al. (2015). Universal access to effective antibiotics is essential for tackling antibiotic resistance. J Law Med Ethics. 43(S3):17–21.

Emmanuel J, Stringer R (2007). For proper proposal: A global inventory of alternative medical waste treatment technologies. Arlington, VA: Healthcare without Harm. (https://noharm-uscanada.org/sites/default/files/documents-files/2046/For_Proper_Disposal.pdf, accessed 06 September 2018).

EPHA, Changing Markets, Alliance to Save Our Antibiotics, and Health Action International (2016). Drug resistance through the back door: How the pharmaceutical industry is fuelling the rise of superbugs through pollution in its supply chains. (http://changingmarkets.org/ wp-content/uploads/2017/05/Drug_resistance_backdoor_final.pdf, accessed 06 September 2018).

European Centre for Disease Prevention and Control, European Medicines Agency (2009). The bacterial challenge: time to react. Stockholm: ECDC/EMEA. (https://ecdc.europa.eu/sites/portal/files/media/en/publications/Publications/0909_TER_The_Bacterial_Challenge_Time_to_React.pdf. accessed 06 September 2018).

Forum on Microbial Threats, Board on Global Health, Institute of Medicine (2006). Ending the war metaphor: The changing agenda for unraveling the host-microbe relationship. Washington, DC: National Academy of

Sciences. (https://www.nap.edu/catalog/11669/ending-the-war-metaphor-the-changing-agenda-for-unraveling-the, accessed 06 September 2018).

Freire-Moran L, Aronsson B, Manz C et al. (2011). Critical shortage of new antibiotics in development against multidrug-resistant bacteria – Time to react is now. Drug Resist Updat. 14(2):118–124.

Friends of the Earth (2017). Chain reaction III: How top restaurants rate on reducing use of antibiotics in their meat supply, 27 September 2017. Washington, DC: Friends of the Earth. (https://foe.org/resources/chain-reaction-iii-report/, accessed 20 January 2019).

Health Care Without Harm (2015). Clinician champions in comprehensive antibiotic stewardship. Virginia: Health Care Without Harm. (https://noharm-uscanada.org/CCCAS, accessed 06 September 2018).

Hopkinson J (2014). FDA checking into growth promotion marketing. Arlington, VA: Politico. (https://www.politico.com/tipsheets/morning-agriculture/2014/06/nyc-soda-limit-case-on-tap-fda-checking-into growth-promotion-marketing-ernst-takes-iowa-212543, accessed 06 September 2018).

Hubbard T, Love J (2004). A new trade framework for global health-care R&D. PLoS Biol. 2(2):147–150.

International Federation of Pharmaceutical Manufacturers & Associations (IFPMA) (2016a). Declaration by the pharmaceutical, biotechnology and diagnostics industries on combating antimicrobial resistance. Geneva: International Federation of Pharmaceutical Manufacturers & Associations. (https://www.ifpma.org/wp-content/uploads/2016/01/briefing-declaration-a4-2017_SCF.pdf, accessed 06 September 2018).

International Federation of Pharmaceutical Manufacturers & Associations (IFPMA) (2016b). Industry roadmap for progress on combating antimicrobial resistance. Geneva: International Federation of Pharmaceutical Manufacturers & Associations. (https://www.ifpma.org/wp-content/uploads/2016/09/Roadmap-for-Progress-on-AMR-FINAL.pdf, accessed 06 September 2018).

Keep Antibiotics Working (2014). Keep Antibiotics Working Coalition disappointed in long awaited antibiotics report from President's Council of Advisors on Science and Technology (PCAST). London: Keep Antibiotics Working. (https://static1.squarespace.com/static/5519650ce4b01b71131cb5f9/t/552ff3ede4b092173bba4c39/1429205997667/KAW+Press+Release+PCAST+Report+9+18+14. pdf, accessed 06 September 2018).

Kunin CM (1993). Resistance to antimicrobial drugs – a worldwide calamity. Annal Intern Med. 118(7):557–561.

Laxminarayan R, Bhutta ZA (2016). Antimicrobial resistance – a threat to neonate survival. Lancet Glob Health. 4(10):e676-e677.

Laxminarayan R, Paudel S, Grigoras C et al. (2016). Access to effective antimicrobials: a worldwide challenge. Lancet. 387(10014):168–175.

Levy SB (2010). Responding to the global antibiotic resistance crisis: the APUA chapter network. In Choffnes E, Relman DA, Mack A eds. Antibiotic resistance: Implications for global health and novel intervention strategies: workshop summary. Washington, DC:National Academies Press.

Mack A, Relman DA, Choffnes ER (2011). Antibiotic resistance: Implications for global health and novel intervention strategies: workshop summary. Washington, DC:National Academies Press.

McDonald's (2017). Vision for antimicrobial stewardship for food animals. McDonald's. (http://corporate.mcdonalds.com/mcd/sustainability/ sourcing/ animal-health-and-welfare/issues-we-re-focusing-on/visionfor-antimicrobial-stewardship-for-food-animals.html, accessed 06 September 2018).

Médecins sans Frontières (n.d.). A Fair Shot campaign. Geneva: Médecins sans Frontières. (https://www.afairshot.org/, accessed 06 September 2018).

Médecins sans Frontières (2016). MSF briefing note: The review on antimicrobial resistance: tackling drug resistant infections globally. Geneva: Médecins sans Frontières. (https://msfaccess.org/sites/default/ files/AMR_MSF_analysis_Oneil.pdf, accessed 06 September 2018).

Molchan S, Rickert J, Powers J (2015). The 21st Century Cures act: More homework to do. Health Affairs Blog. (https://www.healthaffairs.org/do/10.1377/hblog20150924.050749/full/, accessed 06 September 2018).

National Research Defense Council (NRDC) (2015). Press release: As Subway moves to reduce antibiotics in chicken, groups call for details and timelines. New York: National Research Defense Council. (https://www.nrdc.org/media/2015/150901, accessed 06 September 2018).

O'Neill J (2016). Tackling drug-resistant infections globally: Final report and recommendations. London: The Review on Antimicrobial Resistance, p. 7. London: Wellcome Trust and Government of the United Kingdom. (https://amr-review.org/sites/default/files/160518_ Final%20paper_with%20cover .pdf, accessed 06 September 2018).

ReAct (n.d.). Reimagining resistance. Cuenca, Ecuador: ReAct Latin America. (https://www.reactgroup.org/toolbox/raise-awareness/examples-from-the-field/reimagining-resistance/, accessed 06 September 2018).

ReAct (2017). ReAct withdraws from IMI project DRIVE-AB. News and views, news and opinions – 2017. Uppsala: ReAct Group. (https://www .reactgroup.org/news-and-views/news-and-opinions/year-2017/react-withdraws-from-imi-project-drive-ab/, accessed 06 September 2018).

Sanjuan JR (2017). 70th World health assembly intervention, agenda item 12.2: Antimicrobial resistance (AMR). Geneva: Médecins sans Frontières. (https:// www.msfaccess.org/content/70th-world-health-assembly-intervention-agenda-item-122-antimicrobial-resistanceamr, accessed 06 September 2018).

Seabury S, Sood N (2017). Toward a new model for promoting the development of antimicrobial drugs. Health Affairs Blog. (https:// www.healthaffairs.org/ do/10.1377/hblog20170518.060144/full/, accessed 06 September 2018).

Sivaraman S, Murray M (2015). Microbes and Metaphors: A dialogue between scientists, artists and activists. Part of the Reimagining Resistance Series. Uppsala: ReAct Group. (https://www.reactgroup.org/wp-content/ uploads/2016/10/microbes-and-metaphores.pdf, accessed 06 September 2018).

So AD (2014). Welcome presentation: Civil society coalition to tackle antibiotic resistance. Geneva, Switzerland, 28 April–1 May 2014. (https://www .southcentre.int/question/civil-society-workshop-sets-up-new-coalition-on-antibiotic-resistance/, accessed 06 September 2018).

So AD, Shah TA (2014). New business models for antibiotic innovation. Ups J Med Sci. 199(2):176–180.

So AD, Weissman R (2012). Generating antibiotic incentives now: GAIN – or just greed. Huffington Post, 6 June. (https://www.huffingtonpost.com/ entry/antibiotic-resistance-_b_1572284.html?guccounter=1, accessed 06 September 2018).

So AD, Gupta N, Brahmachari SK et al. (2011). Towards new business models for R&D for novel antibiotics. Drug Resist Updat. 14(2):88–94.

So AD, Ramachandran R, Zorzet A et al. (2017). Bridging the gap: A policy briefing on next steps for tackling antimicrobial resistance. Briefing for the World Health Organization Executive Board, EB140/11 140th session. Uppsala: ReAct Group. (https://www.reactgroup.org/wp-content/ uploads/2017/01/ReAct_WHO-EB-Briefing_Jan-2017_LONG.pdf, accessed 06 September 2018).

Spellberg B, Powers JH, Brass EP et al. (2004). Trends in antimicrobial drug development: Implications for the future. Clin Infect Dis. 38(9):1279–1286.

Stern S, Chorzelski S, Franken L et al. (2017). Breaking through the wall: A call for concerted action on antibiotics research and development. Berlin: Boston Consulting Group. (https://www

.bundesgesundheitsministerium.de/fileadmin/Dateien/5_Publikationen/ Gesundheit/Berichte/GUARD_Follow_Up_Report_Full_Report_ final. pdf, accessed 06 September 2018).

Sversut LV (2016). Permanent mission of Brazil to the United Nations Office and other international organizations in Geneva. WIPO, WHO, WTO Joint Technical Symposium on AMR: Panel 3 – Trade policy in support of antimicrobial access and stewardship. (http://www.wipo.int/edocs/ mdocs/mdocs/en/wipo_who_wto_ip_ge_16/wipo_who_wto_ip_ge_16_ www_356197.pdf, accessed 06 September 2018).

UNICEF (2012). Pneumonia and diarrhea: Tacking the deadliest diseases for the world's poorest children. New York: UNICEF. (https://www.unicef .org/publications/index_65491.html, accessed 06 September 2018).

US Department of Agriculture (n.d.). Certified responsible antibiotic use. Washington, DC: US Department of Agriculture. (https://www.ams.usda .gov/services/auditing/crau, accessed 06 September 2018).

USDA Foreign Agricultural Service (2011). Republic of Korea phases out antibiotic usage in compound feed. GAIN, Report Number KS1128. Washington, DC: USDA Foreign Agricultural Service. (http:// battlesuperbugs .com/sites/battlesuperbugs.com/files/CRAU%20 Rationale%20and%20 Standard%20July%202017.pdf, accessed 06 September 2018).

US Food and Drug Administration (2013). FDA update on animal pharmaceutical industry response to guidance #213. Washington, DC: US Food and Drug Administration. (https://www.fda.gov/animalveterinary/ safetyhealth/antimicrobialresistance/judicioususeofantimicrobials/ ucm390738.htm, accessed 06 September 2018).

US Food and Drug Administration (2016). 2015 Summary report on antimicrobials sold or distributed for use in food-producing animals. Washington, DC: US Department of Health and Human Services. (https://www .fda.gov/downloads/ForIndustry/UserFees/ AnimalDrugUserFeeActADUFA/ UCM534243.pdf, accessed 06 September 2018).

US President's Council of Advisors on Science and Technology. (2014). Report to the President on combating antibiotic resistance. Washington, DC: Executive Office of the President. (https://www.cdc.gov/drugresistance/ pdf/report-to-the-president-on-combating-antibiotic-resistance.pdf, accessed 06 September 2018).

US Public Interest Research Group (n.d.). Stop the overuse of antibiotics on factory farms. Denver, CO: US Public Interest Research Group (https:// uspirg.org/issues/usp/stop-overuse-antibiotics-factoryfarms-0, accessed 06 September 2018).

Van Boeckel TP, Brower C, Gilbert M et al. (2015). Global trends in antimicrobial use in food animals. Proc Natl Acad Sci USA. 112(18):5649–5654.

World Economic Forum (2013). Global risks 2013, 8th edn. Geneva: World Economic Forum. (http://www3.weforum.org/docs/WEF_ GlobalRisks_ Report_2013.pdf, accessed 06 September 2018).

World Health Organization (n.d.). Investing in the development of new antibiotics and their consumption: Setting up a global antibiotic research and development partnership. Geneva: World Health Organization. (http:// www.who.int/phi/implementation/consultation_imnadp/en/, accessed 06 September 2018).

World Health Organization (1998). Emerging and other communicable diseases, antimicrobial resistance. Geneva: World Health Organization. (http://apps.who.int/gb/archive/pdf_files/WHA51/ea9.pdf, accessed 06 September 2018).

World Health Organization (2001). WHO global strategy for containment of antimicrobial resistance. Geneva: World Health Organization. (http:// www.who.int/drugresistance/WHO_Global_Strategy_English. pdf, accessed 06 September 2018).

World Health Organization (2005). WHO global principles for the containment of antimicrobial resistance in animals intended for food. Geneva: World Health Organization. (http://apps.who.int/ iris/bitstream/10665/68931/1/ WHO_CDS_CSR_APH_2000.4.pdf, accessed 06 September 2018).

World Health Organization (2012a). The evolving threat of antimicrobial resistance: options for action. Geneva: World Health Organization. (http:// apps.who.int/iris/bitstream/10665/44812/1/9789241503181_eng.pdf, accessed 06 September 2018).

World Health Organization (2012b). Research and development to meet health needs in developing countries: strengthening global financing and coordination. Report of the consultative expert working group on research and development: financing and coordination. Geneva: World Health Organization. (http:// apps.who.int/iris/bitstream/10665/254706/1/9789241503457-eng.pdf. accessed 06 September 2018).

World Health Organization (2015). Global action plan on antimicrobial resistance. Geneva: World Health Organization. (http://apps.who. int/gb/ ebwha/pdf_files/WHA68/A68_R7-en.pdf?ua=1, accessed 06 September 2018).

World Health Organization and Health Action International (2008). Measuring medicine prices, availability, affordability and price components. Geneva:

World Health Organization. (http://www. who.int/medicines/areas/access/ OMS_Medicine_prices.pdf?ua=1, accessed 06 September 2018).

World Health Organization, Sabine Dittrich (FIND), Médecins sans Frontières, and ReAct (2015). Meeting of experts on biomarkers to discriminate bacterial from other infectious causes of acute fever. Geneva, 22–23 September 2015. (https://www.reactgroup.org/uploads/Report_fever%20 biomarker_.pdf, accessed 06 September 2018).

Zuraw L (2014). Antibiotics coalition wants drug companies to stop growth promotion marketing. Food Safety News, 3 November. (https://www .foodsafetynews.com/2014/11/antibiotics-coalition-wants-drug-companies-to-stop-growth-promotion-marketing/, accessed 06 September 2018).

Index

Index note: page numbers in *italics* denote figures or illustrations.

ABS *see* antibiotic stewardship (ABS)
Achaogen, company 191
Acinetobacter 28
   A. baumannii 74, 126, 166, 192
Action on Antibiotic Resistance
   (ReAct) 209, 210, 214, 215,
   217, 230
advertising 207, 212
Africa CDC AMR Surveillance
   network (AMRSNET) 169
Africa Centres for Disease Control and
   Prevention (Africa CDC) 169
AGP (antimicrobial growth promoters)
   101, 103, 106, 111, 224
agricultural sector 37, 101, 223 *see*
   also animal husbandry; farming;
   livestock production
Alliance for Prudent Use of Antibiotics
   (APUA) 208
aminoglycosides 27
amoxicillin 61, 190
ampicillin 190
animal feed 108, 116, 117, 227
animal husbandry 102, 103, 107
   see also farming; livestock
   production
animal products raised without routine
   use of antibiotics 15, 222, 225
animal to human transmission of
   resistant bacteria *104*, 111, 185
Antibiotic Guardian Campaign 60
Antibiotic Resistance Coalition (ARC)
   15, 209, 228
antibiotic stewardship (ABS) 8, 30, 31,
   212, 221 *see also* interventions
   to tackle antimicrobial
   resistance

best practice 87
cost-effectiveness 89, 92
evidence of effectiveness 83
guidelines 86, 208
in hospitals 81
methodology used in studies 85, 92
antibiotic use, in humans
effect of culture 62
nonprescription 4
not targeted 156
reduced, due to vaccination 182,
   186, 195
   economic benefits 196, 198, 200
antibiotic use, in livestock production
   101, 210, 212, 223
in animal feed 108, 227
decrease in 103, 105, 107
growth promoters 101, 103, 106,
   111, 224
interventions to reduce use 114,
   208, 209
measuring 107
risk assessment 110
antibiotics
access to 6, 219
alternative therapies 129
broad spectrum 23
commercialization of new 132, 144
ionophore 102
last-line 2, 162, 168, 221
market approval of 131, 143
pipeline for new 4, 125, 126, 141,
   215
post-antibiotic era 2, 4, 209
promotion and marketing of 208,
   212
prophylaxis 25, 101, 103

antibiotics (cont.)
    research and development 10, 125,
        214
        delinkage 11, 132, 145, 216, 217
        funding 129, 215
        incentives for 11, 132, 212, 217
        see also market entry rewards
            (MERs); prizes
    sales of 132, 144
    second-line 33
    sustainability and systems thinking
        228
    used in both humans and animals
        111, 222
    wastewater contamination with 228
antibodies 129, 191
antigens 186
antimicrobial growth promoters (AGP)
    101, 103, 106, 111, 224
Antimicrobial Resistance Diagnostic
    Challenge 139
APUA (Alliance for Prudent Use of
    Antibiotics) 208
aquaculture 103, 108, 186
ARC (Antibiotic Resistance Coalition)
    15, 209, 228
artemisinin 6
artists 230
ASP (antibiotic stewardship
    programme) see antibiotic
    stewardship (ABS)
authorisation of new antimicrobials
    140
autoimmune diseases 191
avilamycin 111
avoparcin 111
awareness campaigns 54
    Antibiotic Guardian 60
    public health 8, 47
azithromycin 168

bacteria
    eradication of 187
    Gram-negative 2, 126, 128, 131
bacterial infections
    bacteraemia 24
    secondary 195
bacteriophages 129
BARDA (Biomedical Advanced
    Research and Development
    Authority) 138

Bergström, Richard 215
beta-lactams 27
Bezlotoxumab 191
Bill and Melinda Gates Foundation
    139, 146, 191
biomarkers 157, 159, 218
Biomedical Advanced Research
    and Development Authority
    (BARDA) 138
biosecurity measures 103, 107, 110,
    114, 116
biosurveillance see surveillance
    programmes
bloodstream infections (BSIs) 30, 71,
    73, 74, 89
booklets, for patients 58, 59
broad spectrum antibiotics 23
budgets, health care 32
Burden of Resistance and Disease in
    European Nations (BURDEN)
    71
bystander effects 182, 195, 196

campaigns 54
    Antibiotic Guardian 60
    public health 8, 47
Campylobacter 28, 127
cancer treatments 25, 191
Candida 28
carbapenems 2, 27, 31, 192
    resistance 28, 126
CARB-X (Combating Antibiotic
    Resistant Bacteria
    Biopharmaceutical Accelerator)
    136
ceftriaxone 162, 168
cephalosporins 5, 31, 72, 72
Chan, Margaret 209
chemotherapy 25
chicken farming 102, 116, 117
children 59, 195, 222
chloroquine 6
cholera 190
ciprofloxacin 162, 193, 194
civil society 15, 207
cleaning, in hospitals 74
clinical trials 11, 131, 136, 142, 170
Clostridium difficile 191
Clostridium perfringens 102
coccidiosis 102
colistin 2, 111, 192

colony-forming units 77
Combating Antibiotic Resistant
    Bacteria Biopharmaceutical
    Accelerator (CARB-X) 136
commercialization of new antibiotics
    132, 144
communication, with patients 53, 58,
    59, 63
community, prescriptions for
    antibiotics 7
companion animals 113
cooperation, global 15, 16, 146
cost
    of diagnostic research and
        development 171
    of interventions 61
    of market entry rewards 145
    societal, of antimicrobial resistance
        34
    of vaccinations 198
cost-effectiveness
    of antibiotic stewardship in hospitals
        89, 92
    of biosecurity in livestock
        production 115, 116
    of infection control measures in
        hospitals 74
    of preventative strategies in the
        community 38, 61
cough 58, 59 *see also* lower respiratory
    tract infections
C-reactive protein (CRP) 57, 58,
    61, 157
cross transmission of infection 77
Cross-Research Council AMR
    Initiative 139
cross-species resistance 210
culture, and antibiotic use 62

deaths, due to antimicrobial resistance
    24, 25, 35, 222
decolonization 74
delayed prescribing (DP) 54, 57
delinkage 11, 132, 145, 216, 217
dentists 46
    development pipeline for new
        antibiotics 4, 125, 126,
        141, 215 *see also* research
        and development into new
        antibiotics
diagnostic tests 12, 89, 218

for antimicrobial resistance
    surveillance 164
barriers to innovation 162
business case for 174
cost 61, 171
decrease cost of clinical trials 170
efficient implementation 173
funding for 171, 172
for malaria 6
pathogen detection 160
point-of-care 52, 57, 58, 144, 156
policies 172
to reduce misuse of antibiotics 164
regulatory approval 171
susceptibility testing 160
using host biomarkers 157
diarrhoea
    in children 193, 222
    in piglets 102, 116
    vaccines 196
Directorate-General for Research and
    Innovation (DG-RTD) 137
disability affected life years (DALYs) 25
disease prevention 105, 107, 110
    role of diagnostic tests 174
doctors (GPs) 46, 58
Driving Reinvestment in Research and
    Development and Responsible
    Antibiotic Use (DRIVE-AB) 81,
    219
Drugs for Neglected Diseases Initiative
    (DNDi) 218

ear infections 34
EARS-Net (European Antimicrobial
    Resistance Surveillance
    Network) 71, 78, 79, 168
ECDC (European Centre for Disease
    Prevention and Control) 71, 87,
    109, 110
economics
    of antibiotic research 132, 133
    benefits of antibiotic stewardship 92
    benefits of vaccines 196, 198, 200
    of biosecurity 116
    burden of antimicrobial resistance 6,
        31, 35, 38
    health 92
EDCTP (European and Developing
    Countries Clinical Trial
    Partnership) 136

education
  of clinicians 51, 58
  of patients 53, 58, 173
  of undergraduates 91
Effective Practice and Organisation of
    Care Group (EPOC) 81
EFPIA (European Federation of
    Pharmaceutical Industries and
    Associations) 137
EFSA (European Food Safety
    Authority) 109
EMA (European Medicines Agency)
    140, 144
empirical treatment 23
enteritis, necrotic, in poultry 102, 117
Enter-net 168
Enterobacteriaceae 24, 31, 72, 74, 192
  ESBL (extended spectrum beta-
    lactamase) 113
Enterococcus faecium 127
environment *see also* One Health
    issues
  antibiotics entering 210, 228
  transmission of resistant bacteria
    104, 111
environmental cleaning, in hospitals 74
EPOC (Effective Practice and
    Organisation of Care Group) 81
EQUIP project 59
eradication, of bacteria 187
ESAC-Net (European Surveillance of
    Antimicrobial Consumption
    Network) 71
ESBL (extended spectrum beta-
    lactamase) 113, 115
Escherichia coli 2, 31, 33, 165
  cephalosporin resistant *5*, 36, 72
  colistin-resistant 111
  enterotoxigenic *E. coli* (ETEC) 193
  extended spectrum beta-lactamase
    (ESBL) 115
  uropathogenic *E. coli* (UPEC) 194
  vaccines against 186, 193
  Verocytotoxin-producing (VTEC)
    168
Essential Medicines List 207
essential oils 116
ESVAC (European Surveillance of
    Veterinary Antimicrobial
    Consumption) 108, 109

ETVAX, vaccine 193
EU (European Union) 137
  One Health Action Plan ix
European and Developing Countries
    Clinical Trial Partnership
    (EDCTP) 136
European Antimicrobial Resistance
    Surveillance Network (EARS-
    Net) 71, 78, 79, 168
European Awareness Day 54
European Centre for Disease
    Prevention and Control (ECDC)
    71, 87, 109, 110
European Federation of
    Pharmaceutical Industries and
    Associations (EFPIA) 137
European Food Safety Authority
    (EFSA) 109
European Medicines Agency (EMA)
    140, 144
European Surveillance of
    Antimicrobial Consumption
    Network (ESAC-Net) 71
European Surveillance of Veterinary
    Antimicrobial Consumption
    (ESVAC) 108, 109
European Union *see* EU
extended spectrum beta-lactamase
    (ESBL) 113, 115
extrinsic resistance 27

farmers 113, 116
farming *see also* agricultural sector;
    animal husbandry; livestock
    production
  chicken 102, 116, 117
  intensive 10, 102, 112, 223
  pig 102, 114, 116
  salmon 186
fatality rates 24, 25, 35, 222
FDA (US Food and Drug
    Administration) 140, 144
feed, animal 108, 116, 117, 227
feedback 51
financial incentives 52
FIND (Foundation for Innovative New
    Diagnostics) 157, 218
flavomycin 111
Fleming, Sir Alexander 2
fluoroquinolones 27, 195

food
  animal feed 108, 116, 117, 227
  foodborne infections 185
  hygiene 113
  transmission of resistance 113, 185
Food and Drug Administration (FDA)
    (US) 140, 144
food animal production *see also*
    livestock production
  use of antibiotics in 9, 223
  without routine use of antibiotics
    15, 222, 225
Foodborne Diseases Active Surveillance
    Network (FoodNet) 169
Foundation for Innovative New
    Diagnostics (FIND) 157, 218
funding
  for antibiotic research and
    development 129, 215
  for diagnostic tests 171, 172
furunculosis 186

G3REC (*E. coli* resistant to 3rd
    generation cephalosporins) 72
G20 summit, Hamburg 2017 146
GAMRIF (Global Antimicrobial
    Resistance Innovation Fund) 136
GARDP (Global Antibiotic Research
    and Development Partnership)
    135, 218
GARP (Global Antibiotic Resistance
    Partnership) 6
GASP (Global Gonococcal
    Antimicrobial Surveillance
    Programme) 168
Gavi Vaccine Alliance 172, 194, 198
GDP (gross domestic product) 37
general practitioners (GPs) 46, 58
GLASS (Global Antimicrobial
    Resistance Surveillance System)
    165
Global Action Plan on
    Antimicrobial Resistance
    (WHO) 3, 15, 105, 155, 164,
    212, 213, 221
Global Antibiotic Research and
    Development Partnership
    (GARDP) 135, 218
Global Antibiotic Resistance
    Partnership (GARP) 6

Global Antimicrobial Resistance
    Collaboration Hub 12, 16, 146
Global Antimicrobial Resistance
    Innovation Fund (GAMRIF)
    136
Global Antimicrobial Resistance
    Surveillance System (GLASS)
    165
Global Challenge Research Fund 139
global cooperation 15, 16, 146
Global Gonococcal Antimicrobial
    Surveillance Programme (GASP)
    (WHO) 168
Global Point Prevalence Survey of
    Antimicrobial Consumption
    and Resistance (GLOBAL-PPS)
    169
Global Strategy for Containment of
    Antimicrobial Resistance 208
gonococcal resistance 160, 168
gonorrhoea 160, 161, 168
governing body, global 146.2
GPs (general practitioners) 46, 58
GRACE INTRO project 58
Gram-negative bacteria 2, 126, 128,
    131
grants, research 132, 142
gross domestic product (GDP) 37
growth promoters, antimicrobial 101,
    103, 106, 111, 224
guidelines 51, 86, 88

Haemophilus influenzae 127, 187
  vaccines 190
hand hygiene 73, 77
health
  campaigns 8, 47
  economics 92
  One Health issues ix, 15, 111, 209,
    223, 228
Health Action International (HAI)
    207, 209, 212
health burden, of antimicrobial
    resistance 24, 27
health care budgets 32
health care-associated infections
    (HAIs) 8, 71, 72
  outbreak control 79
  surveillance programmes 71, 78,
    90, 91

health technology assessment (HTA)
13, 172
Healthcare without Harm 222, 226,
229
Helicobacter pylori 127
Heymann, Dr. David 209
HNL (human neutrophil lipocalin) 159
Horizon 2020 Better Use of Antibiotics
Prize 137, 159
hospitals *see also* health care-
associated infections (HAIs)
antimicrobial resistance in 8, 113
blocked beds 76
environmental cleaning 74
overcrowding 77
vaccines and immunotherapies for
resistant bacteria 191
waste management 229
host biomarkers 157, 159, 218
host-microbe relationship 230
HTA (health technology assessment)
13, 172, 172
human neutrophil lipocalin (HNL) 159
human to animal transmission of
resistant bacteria *104*, 111, 185
husbandry, animal 102, 103, 107
see also farming; livestock
production
hygiene
cleaning, in hospitals 74
food 113
hand 73, 77

IMI (Innovative Medicine's Initiative)
137
immune stimulation 129
immunocompromized patients 25
immunotherapies 191 *see also*
monoclonal antibodies (mAbs);
vaccines
incentives for antibiotic research and
development 11, 132, 212, 217
see also market entry rewards
(MERs)
incidence of antimicrobial resistance 29
infection prevention and control (IPC)
8, 25, 30, 72
cost-effectiveness 74
infection control measures 73
prevention of cross transmission 77
by vaccines 182

infections
bacterial 24, 195
bloodstream (BSIs) 30, 71, 73, 74,
89
ear 34.1
health care-associated infections
(HAIs) 8, 71, 72
outbreak control 79
surveillance programmes 71, 78,
90, 91
incidence of 26
lower respiratory tract (LRTIs) 58
reduced severity of 182
respiratory tract infections (RTIs)
14, 46, 52, 59
urinary tract (UTIs) 61, 194
viral 46, 52, 54, 57, 195
influenza 182
vaccines 195
Innovative Medicine's Initiative (IMI)
137
InnovFin ID (Infectious Diseases
Facility) 138
intensive animal production 10, 102,
112, 223
international response to antimicrobial
resistance 15, 16, 146
international standards, for diagnostic
tests 172
interspecific effects 184
interventions to tackle antimicrobial
resistance *see also* antibiotic
stewardship (ABS)
clinician and patient focused 53
clinician focused 51
evidence of effectiveness 83
in hospitals 81
long-term impact 63
methodology used in studies 85, 92
projects 58, 59
public focused 54
intrinsic resistance 27
ionophore antibiotics 102
IPC *see* infection prevention and
control (IPC)
isolation 73
Italy, University hospital Modena 87

Japanese Pharmaceuticals and Medical
Devices Agency (PMDA) 141,
144

Joint Programming Initiative on
    Antimicrobial Resistance
    (JPIAMR) 135

KFC, restaurant chain 225
Klebsiella pneumoniae 2, 27, 31, 166,
    192
  cephalosporin resistant 5, 36
  potential vaccine 199
Korea, Republic of 227
Krankenhaus-Infektions-Surveillance-
    System (KISS) 78

last-line antibiotics 2, 162, 168, 221
legal framework, international 16
Limited Population Antibacterial Drug
    (LPAD) 140
livestock production 9, 37 *see also*
    agricultural sector; animal
    husbandry; farming
  animals raised without routine use
    of antibiotics 15, 222, 225
  antibiotic use in 101, 210, 212,
    223
    decrease 103, 105, 107
    measuring 107
    risk assessment 110
  growth promoters, antimicrobial
    101, 103, 106, 111, 224
  intensive 10, 102, 112, 223
  interventions to reduce antimicrobial
    use 114, 208, 209
  productivity 105
  transmission of resistant bacteria to
    humans and the environment
    104, 111, 185
livestock-associated methicillin-
    resistant *Staphylococcus aureus*
    (LA-MRSA) 113
Longitude Prize 159
low and middle-income countries
    (LMICs) 4, 8, 9, 139
lower respiratory tract infections
    (LRTIs) 58
LPAD (Limited Population
    Antibacterial Drug) 140
lysins 129

malaria 6, 36, 182
manufacturing plants 228

market approval of antibiotics 131,
    143
market entry rewards (MERs) 11, 144,
    217 *see also* prizes
marketing, of antibiotics 208, 212
McDonald's, restaurant chain 226
meat, produced without antibiotics 15,
    222, 223
Médecins sans Frontiers (MSF) 209,
    217, 218
medicines
  ban on advertising 207
  Essential Medicines List 207
meningitis 190
metaphylactic antimicrobial use 99,
    101, 102
methicillin-resistant *Staphylococcus
    aureus* (MRSA) 26, 28, 36, 72
  assays for 167
  livestock-associated (LA-MRSA)
    113
microbe-host relationship 230
Modena, University hospital 87
monoclonal antibodies (mAbs) 191 *see
    also* immunotherapies
mortality, major causes of 192
mortality rates 24, 25, 35, 222
MRSA 26, 28, 36, 72
  assays for 167
  livestock-associated (LA-MRSA) 113
MSF (Médecins sans Frontiers) 209,
    217, 218

National Institute for Health Research
    (NIHR) 139
National Institute of Allergy and
    Infectious Diseases (NIAID) 138
ND4BB (New Drugs for Bad Bugs)
    137
necrotic enteritis, in poultry 102, 117
Neisseria gonorrhoeae 28, *127*, 166,
    168
Neisseria meningitidis 187, 190
Netherlands 88
neutrophil biomarkers 159
New Drugs for Bad Bugs (ND4BB)
    137
Newton Fund 139
NIAID (National Institute of Allergy
    and Infectious Diseases) 138

NIHR (National Institute for Health
    Research) 139
norovirus 196
nosocomial bacterial pathogens 191
Nunan, Cóilín 224
nurses 46

OIE (World Organisation for Animal
    Health) 99, 108, 185, 214
One Health issues ix, 15, 111, 209,
    223, 228
online guidelines 88
opportunity cost 38
outbreaks of health care-associated
    infections 79
overcrowding, in hospitals 77

parents 59
patients 47, 92, 212
    educational materials for 53, 58, 59
peptides 129
pets 113
pharmaceutical industry 207, 210
Pharmaceuticals and Medical Devices
    Agency (PMDA), Japan 141,
    144
pharmacists 46
pig farming 102, 114, 116
pipeline for new antibiotics 4, 125,
    126, 141, 215 *see also* research
    and development into new
    antibiotics
Plasmodium falciparum 182
pledges, to reduce antimicrobial
    resistance 60
PMDA (Pharmaceuticals and Medical
    Devices Agency), Japan 141,
    144
pneumonia 218, 222
    pneumococcal conjugate vaccine
    14, 218
point prevalence studies 167, 169
point-of-care diagnostic tests (POCTs)
    52, 58, 144, 156
    for antimicrobial resistance
    surveillance 164
    barriers to innovation 162
    business case for 174
    cost 61, 171
    decrease cost of clinical trials 170

efficient implementation 173
    funding for 172
    pathogen detection 160
    policies 172
    to reduce misuse of antibiotics 57,
    164
    regulatory approval 171
    susceptibility testing 160
    using host biomarkers 157
pollution, antibiotic 228
polymyxins 31
post-antibiotic era 2, 4, 209
poultry, necrotic enteritis 102, 117 *see
    also* chicken farming
prebiotics 116
preclinical trials 131, 142
prescribing, delayed 54, 57
prescription medicines, ban on
    advertising 207
prescriptions, for antibiotics 7, 46, 61,
    108
    audit and feedback 51
preventative strategies 39
priority pathogens list (PPL) 126
prizes 137, 159 *see also* market entry
    rewards (MERs)
probiotics 116, 129
procalcitonin 52, 57, 157
productivity, in livestock sector 105
profits, from antibiotic sales 144
projects to tackle antimicrobial
    resistance
    EQUIP 59
    GRACE INTRO 58
prophylactic antimicrobial use 25,
    100, 101, 103
Pseudomonas aeruginosa 28, 74, 126,
    128, 192
    monoclonal antibodies 191
    vaccine 191
public health campaigns 8, 47

Qualified Infectious Diseases Products
    (QIDP) 140

ReAct (Action on Antibiotic
    Resistance) 209, 210, 214, 215,
    217, 230
regulatory approval for diagnostic tests
    171

regulatory initiatives for new
   antimicrobials 140, 143, 220
reminders 51, 58
Republic of Korea 227
research and development into new
   antibiotics 10. 125, 214
  delinkage 11, 132, 145, 216, 217
  funding 129, 215
  incentives for 11, 132, 212, 217
     see also market entry rewards
     (MERs)
  pipeline 4, 125, 126, 141, 215
research grants 132, 142
resistance
  extrinsic 27
  genes, transfer of 185
  intrinsic 27
respiratory syncytial virus (RSV) 195
respiratory tract infections (RTIs) 14,
   46, 52, 59
  lower respiratory tract (LRTIs) 58
restaurant chains 15, 225
rotavirus vaccines 196

safety, patient 92, 212
sales, of antibiotics 132, 144
salmon farming 186
Salmonella 116, 168
Salmonella spp. 127, 166
  S. typhi, vaccines for 194
Schippers, Edith 227
screening 74
secondary bacterial infections 195
second-line antibiotics 33
self-care 53
self-limiting symptoms 54
shared decision-making (SDM) 53, 55
Shigella spp. 127, 166
smallpox 187
societal costs of antimicrobial
   resistance 34
society, civil 15, 207
specificity, of vaccines 186
standards
  for diagnostic tests 172
  for licensed drugs 143
Staphylococcus aureus 33, 127, 128,
   166
  methicillin-resistant (MRSA) 26, 28,
   36, 72, 113, 167

monoclonal antibodies 191
  vaccine 191
stewardship *see* antibiotic stewardship
   (ABS)
Streptococcus pneumoniae 28, 127,
   166
  vaccine 187
Subway, restaurant chain 225
surgical procedures 25
surveillance programmes
  antibiotic consumption, in
     agriculture 109
  antibiotic resistance 71, 78, 90, 91,
     109, 212
  in developing countries 167
  diagnostic tests 164
  Africa CDC AMR Surveillance
     network (AMRSNET) 169
  Enter-net 168
  European Antimicrobial Resistance
     Surveillance Network (EARS-
     Net) 71, 78, 79, 168
  Foodborne Diseases Active
     Surveillance Network
     (FoodNet) 169
  Global Gonococcal Antimicrobial
     Surveillance Programme (GASP)
     168
  Global Point Prevalence Survey of
     Antimicrobial Consumption
     and Resistance (GLOBAL-PPS)
     169
surveys 91
sustainability and systems thinking
   228
symptoms, self-limiting 54

therapeutic antimicrobial use 99, 101,
   102, 106, 224
toxicity 111, 187
Transatlantic Task Force on
   Antimicrobial Resistance
   (TATFAR) 141, 144, 215
transmission
  of antimicrobial resistance by water
     113, 185, 228
  of infection in hospitals 77
  of resistant bacteria between
     animals, humans and the
     environment *104*, 111, 185

treatment, empirical 23
trends in antimicrobial resistance rates
31
tuberculosis 34, 36
Ty21a, vaccine 194

UK Research and Innovation 139
understaffing, of hospitals 77
United Kingdom (UK) 139
United Nations (UN) 3, 214
United States (US) 138
antibiotics in food animal
production 224
Food and Drug Administration
(FDA) 140, 144
University hospital Modena,
Italy 87
urinary tract infections (UTIs) 61, 194

Vaccine Alliance, GAVI 172, 194, 198
vaccines 14, 102, 103, 116, 129, 218
see also immunotherapies
advantages 186
in animal production 185
cost 198
difficulties in developing 193
economic benefits 196, 198, 200
impact on antimicrobial
prescriptions 199
reducing antibiotic use 182, 186,
195
reducing antimicrobial resistance
182, 199
specificity of 186
value of 198
for *Clostridium difficile* 191
for *Escherichia coli* 186, 193
for *Haemophilus influenzae* 190
for *Pseudomonas aeruginosa* 191
for rotavirus 196

for *Salmonella typhi* 194
for *Staphylococcus aureus* 191
for *Streptococcus pneumoniae* 187
for urinary tract infections 194
for *Vibrio cholerae* 190
for viruses 195
ETVAX 193
pneumococcal conjugate 218
Ty21a 194
Vi-polysaccharide 194
WC-rBS 193
vancomycin resistance 111
vancomycin-resistant Enterococci
(VRE) 28
Verocytotoxin-producing *E. coli*
(VTEC) 168
veterinary prescriptions 108
Vibrio cholerae, vaccines 190
Vi-polysaccharide, vaccine 194
viral infections 46, 54
diagnostics 52, 57
vaccines for 195

waste
in environment 113
management practices of hospitals
229
wastewater contamination with
antibiotics 228
water, as a means of transmission of
antimicrobial resistance 113,
185, 228
WC-rBS, vaccine 193
websites 60
Wellcome Trust 140, 147
white blood cell count 157
World Health Organization (WHO) 2,
155, 207, 208, 218, 221
World Organisation for Animal Health
(OIE) 99, 108, 185, 214